T0320674

Integrative Bioinformatics for Biomedical Big Data
A No-Boundary Thinking Approach

The volume and complexity of biological and biomedical research continues to grow exponentially with cutting-edge technologies such as high-throughput sequencing. Unfortunately, bioinformatics analysis is often considered only after data have been generated, which significantly limits the ability to make sense of complex big data. This unique book introduces the idea of No-Boundary Thinking (NBT) in biological and biomedical research, which aims to access, integrate, and synthesize data, information, and knowledge from bioinformatics to define important problems and articulate impactful research questions. This interdisciplinary volume brings together a team of bioinformatics specialists who draw on their own experiences with NBT to illustrate the importance of collaborative science. It will help stimulate discussion and application of NBT, and will appeal to all biomedical researchers looking to maximize their use of bioinformatics for making scientific discoveries.

XIUZHEN HUANG is currently Professor and Research Scientist of the Department of Computational Biomedicine at Cedars-Sinai Medical Center, USA. Before joining Cedars-Sinai Medical Center fall 2022, Xiuzhen was Professor in Computer Science at Arkansas State University for eighteen years. She founded the Arkansas Artificial Intelligence Campus (Arkansas AI-Campus). She also initiated the National AI-Campus, a landing project with No-Boundary Thinking. She completed her doctorate degree in computer science at Texas A&M University (2004).

JASON H. MOORE is Chair of the Department of Computational Biomedicine at Cedars-Sinai Medical Center, USA, where he also leads the Center for Artificial Intelligence Research and Education (CAIRE). He is a biomedical informaticist with a focus on AI and machine learning methodology for the study of human health. He has contributed to more than 600 publications in journals and books spanning the clinical and translational sciences, basic sciences, and data science and bioinformatics. He also serves as Editor-in-Chief of the journal *BioData Mining*. He is an elected Fellow of the American College of Medical Informatics and the American Statistical Association.

YU ZHANG is Professor and Chair of the Department of Computer Science at Trinity University, USA. She is Editor-in-Chief of the *International Journal of Agent Technologies and Systems* and is on the editorial board of *SIMULATION: Transactions of The Society for Modeling and Simulation International*. Her research falls within the fields of agent-based modeling and simulation. She has received the Outstanding Service Award of the Society for Modeling and Simulation International (2013), Trinity Distinguish Junior Faculty Award (2008), Best Paper Award of the IEEE Region 5 Student Paper Competition (2008), and the IEEE Central Texas Chapter Service Recognition (2007).

Integrative Bioinformatics for Biomedical Big Data

A No-Boundary Thinking Approach

Edited by

XIUZHEN HUANG
Cedars-Sinai Medical Center, California

JASON H. MOORE
Cedars-Sinai Medical Center, California

YU ZHANG
Trinity University, Texas

CAMBRIDGE
UNIVERSITY PRESS

Shaftesbury Road, Cambridge CB2 8EA, United Kingdom

One Liberty Plaza, 20th Floor, New York, NY 10006, USA

477 Williamstown Road, Port Melbourne, VIC 3207, Australia

314–321, 3rd Floor, Plot 3, Splendor Forum, Jasola District Centre,
New Delhi – 110025, India

103 Penang Road, #05–06/07, Visioncrest Commercial, Singapore 238467

Cambridge University Press is part of Cambridge University Press & Assessment,
a department of the University of Cambridge.

We share the University's mission to contribute to society through the pursuit of
education, learning and research at the highest international levels of excellence.

www.cambridge.org
Information on this title: www.cambridge.org/9781107114302

DOI: 10.1017/9781316335208

© Cambridge University Press & Assessment 2023

First published 2023

A catalogue record for this publication is available from the British Library.

A Cataloging-in-Publication data record for this book is available from the Library
of Congress.

ISBN 978-1-107-11430-2 Hardback

This book is dedicated to all the scientists and future scientists in the world

Contents

Contributors

Clare Bates Congdon
Department of Computer Science, Bowdoin College, Brunswick,
ME, USA

Bryan Dewsbury
Department of Biological Sciences, University of Rhode Island, South
Kingstown, RI, USA

Mariola J. Ferraro
Department of Microbiology and Cell Science, College of Agricultural
and Life Sciences, University of Florida, Gainesville, FL, USA

James A. Foster
Institute for Bioinformatics and Evolutionary Studies (BEST),
University of Idaho, Moscow, ID, USA

Matt Hibbs
Department of Computer Science, Trinity University, San Antonio,
TX, USA

Xiuzhen Huang
Department of Computational Biomedicine, Cedars-Sinai Medical
Center, Los Angeles, CA, USA

Jason H. Moore
Department of Computational Biomedicine, Cedars-Sinai Medical
Center, Los Angeles, CA, USA

Alison Motsinger-Reif
Department of Statistics, North Carolina State University, Raleigh, NC, USA

Bindu Nanduri
Department of Comparative Biomedical Sciences, College of Veterinary Medicine, Mississippi State University, Mississippi State, MS, USA

Joan Peckham
Department of Computer Science and Statistics, University of Rhode Island, South Kingstown, RI, USA

Andy D. Perkins
Department of Computer Science and Engineering, Mississippi State University, Mississippi State, MS, USA

Mahalingam Ramkumar
Department of Computer Science and Engineering, Mississippi State University, Mississippi State, MS, USA

Donald C. Wunsch II
Department of Electrical & Computer Engineering, Missouri University of Science and Technology, Rolla, MO, USA

Yu Zhang
Department of Computer Science, Trinity University, San Antonio, TX, USA

1   No-Boundary Thinking

Xiuzhen Huang and Jason H. Moore

What is No-Boundary Thinking (NBT)? Is it a philosophy term or a science term? Why do we need it? Since 2013, the NBT national network has had many discussions and today wants to have a book to include some of the NBT group members' thoughts. Some may affect NBT, some may not. Still, we would like to put it all together.

What is NBT? No-Boundary Thinking is no-boundary problem defining on time chain to address real science challenges. This is the definition for now, and for many future years in science. No-Boundary Thinking is still No-Boundary Thinking, even with subtraction and more subtraction. Even with addition and more addition, multidisciplinary research is still multidisciplinary. It is different from multidisciplinary, interdisciplinary, or transdisciplinary research. It is also different from the *convergence* approach currently promoted by the National Science Foundation (NSF). No-Boundary Thinking is like the sea: sometimes the sea is rising, sometimes the sea is retiring; it is still the complex sea.

Current NSF/NIH (National Institutes of Health) projects are like collecting water in many confined pools, and even convergence aims to connect these pools into big pools. Of course, they make some contributions to science. However, they are not like running rivers. Running rivers could be big or small, but they lead to the sea, and are eventually parts of the sea.

Today we see the rapid development of science and technology, and the great accumulation of knowledge and wealth. We have powerful machines, high-performance computers, and broadband Internet. We see more and more data being generated, collected, and distributed, and we see new interdisciplinary areas of science appearing and expanding rapidly. Recently we have seen the surge of artificial intelligence, with big data and powerful computers, which is changing the frontiers of science research and industrial innovation. People are expecting new heights in science, a new paradigm.

We know there were big jumps in science and technology throughout human history, with great impacts. Five hundred years ago the Renaissance brought us Da Vinci and Michelangelo, and the new thinking of humanism, which stimulated the development of all areas, including art, science, architecture, politics, and literature. Two hundred years ago the Industrial Revolution brought us Watt, Fulton, and Stephenson, and new machines tools, factory systems, population growth, and wealth accumulation that changed everyday life.

Today, people are proclaiming a new paradigm is coming: the big data paradigm. Data-intensive computing (big data) was advocated as the fourth paradigm for scientific discovery (Hey et al. 2009). In recent years researchers, funding agencies, and companies have promoted data science research. The federal funding agencies – the NSF and NIH – have made large investments in big data and related programs, such as NIH BD2K and NSF BIGDATA. Recently the NSF has identified one of its 10 big ideas, *convergence*, as "a means of solving vexing research problems" and described it as "the closest to transdisciplinary research" (NSF n.d.).

But, we wonder: Can big data, data science, transdisciplinarity, or convergence really bring us what we expect – a new age in science, a new paradigm? Powerful machines drove the Industrial Revolution; can big data similarly drive the current rise of science?

Since 2013 we have been discussing, as a national network, how to address real science challenges and the limitations of big data and interdisciplinary research. We have discussed the future development of bioinformatics and the issues of big data (Huang et al. 2013, 2015; Moore et al. 2017). Big data seems unable to address science challenges the way it promises to, such as the Human Genome Project and The Cancer Genome Atlas (TCGA) project.

In our previous paper (Huang et al. 2015) we discussed the TCGA initiative as an example:

For the pilot project and phase II of TCGA, about US$200-million has been invested in this effort to gather samples, generate data, and analyze the data. The Cancer Genome Atlas (TCGA), may have produced some good results published in Nature and Science, but the approach of big data overall is disconnecting researchers and science challenges. Efforts like TCGA are reaching the "bottleneck;" it is hard to make significant breakthroughs in scientific challenges by focusing on big data and over-simplified problems.

Two years ago I asked one of my collaborators, who worked in the research area of cancer genomics for 30 years and recently retired: Given another 30 years, would you design your research the same way? He directly answered: No.

Current funding agencies, including the NSF and NIH, in a way encourage researchers to focus on oversimplified problems that can be solved or for which results can be generated in a funding period of 3–5 years. However, for many real science challenges we know we may not be able to resolve them in our lifetimes. When we explore science challenges and conduct research design we may consider leaving a window for the next generation to conduct research.

A senior scientist once asked me: Multidisciplinary, interdisciplinary, or transdisciplinary research seems ineffective and cannot address real science challenges – how about "very" interdisciplinary? I asked: But how "very" is enough?

We promote NBT. The intellectual basis of the Renaissance was humanism, where "Man is the measure of all things." No-Boundary Thinking is a science renaissance, where "human intelligence" with NBT is the measure of science.

No-Boundary Thinking is no-boundary problem defining. One intelligent young scientist said his senior advisor told him: "Do not continue to stay in this house to try to do further carving and decorating, or try to add more refined adds-on to the house anymore; too many people have worked on this house. You should leave and go to build a new house." We would suggest to answer: "I am not going to leave, because I want this land. I want to tear down to remove the house here and build a new house here, indeed, a new mansion on this land." This piece of land is the real science challenge; a restructured mansion is the redefined problem, with the problem solution incorporated into the problem definition.

Why time chain? The time chain is a new concept; it is related to the process of problem defining and redefining. It is no-boundary problem defining and redefining. The new mansion may not be as refined as the previous house at the beginning, but its structure is clear and different, and it has the ability to self-recycle and even to self-restructure. As time passes on the time chain, the mansion may all be cleared up one day, and this piece of clear land may be connected with other lands to build a new structure. No-Boundary Thinking, with the land and the time chain, is high-dimensional. NBT is the pursuit of no-boundary thinking in science and in science history.

The purpose of this book is not to present NBT research results or education outcomes, but to stimulate more thought regarding NBT in science, research, and education.

REFERENCES

Hey T, Tansley S, Tolle K 2009. *The Fourth Paradigm: Data-Intensive Scientific Discovery*. Micro Research.

Huang X, Bruce B, Buchan A, et al. 2013. No-boundary thinking in bioinformatics research. *BioData Mining*, 6:19.

Huang X, Jennings SF, Bruce B, et al. 2015. Big data: a 21st century science Maginot Line? No-boundary thinking: shifting from the big data paradigm. *BioData Mining*, 8:7.

Moore JH, Jennings SF, Greene CS, et al. 2017. No-boundary thinking in bioinformatics. *Pac Symp Biocomput*, 22:646–648.

NSF, n.d. Convergence. www.nsf.gov/od/oia/convergence/index.jsp

2 Artificial Intelligence Approaches to No-Boundary Thinking

Jason H. Moore

2.1 INTRODUCTION

A central goal of artificial intelligence (AI) research is to develop computational systems and software that can reason and solve complex problems as humans do. The field of AI has gone through a number of ups and downs, including a period of time during the 1980s and 1990s that was referred to as the "AI winter" because the technology didn't live up to its hype (Crevier 1993). Many believe the AI winter ended in 2010 when the Watson AI software from IBM beat the human champion of a popular television game show (Ferrucci 2012). Watson was a major technological advance in AI research and combined many computational methods – including machine learning, information retrieval, and knowledge representation – to demonstrate human-competitive results on a complex problem. Watson is now being evaluated in the healthcare domain with mixed results (Strickland 2019).

Perhaps the most impactful advance in AI in the last 15 years has been the explosion of methods based on deep learning neural networks (Hinton and Salakhutdinov 2006). Neural networks have been around for decades, but have recently become more useful due to innovative new algorithms and advances in computing hardware such as graphical processing units (GPUs). Their impact in the biomedical domain has been recognized in areas such as pathology and radiology, where deep learning has been useful for the analysis of images. Recent reviews document many of the deep learning successes and some of the challenges (Ching et al. 2018; Topol 2019). An interesting question is whether deep learning is AI. On the surface, deep learning is really a subset of machine learning in that it takes a set of inputs, mathematically transforms those inputs, and produces a set of outputs that meet some objective such as classification. By this definition, deep learning is a component of a sub-discipline of AI. However, deep learning can be used for visual perception, acoustic analysis, and classification, which means it could be pieced together in the context of robotics. This could impact surgery (Esteva et al. 2019), for example, as a human-competitive task that crosses over into the broader AI realm.

An area where AI has tremendous potential is data science. That is, can we build computational systems that are able to perform data analysis as a human does? Human-based data analysis is complex because it involves selecting an analytical method and its parameter settings, performing the analysis, assessing the quality of the results, performing both biological and statistical interpretation of the results, assessing the utility of the result, and then internalizing this entire experience to be used to perform a better data analysis in the future. Data analysis is even more complex than this because humans don't always follow a script. We often try different approaches to see what works and doesn't work before we settle on an analytics strategy. Thus, we as humans "tinker" with data. We also have concepts of "interestingness" that can guide us in our decision-making (Geng and Hamilton 2006). A computer that could do all of this would certainly fall into the realm of AI and would have the potential to greatly advance analytics-based discovery using observational or experimental data.

The idea of letting the computer make the decision and do the analysis has been front and center in a new area of AI called automated machine learning, or AutoML, that has been documented in a recent book (Hutter et al. 2019). The goal of AutoML is to take some of the guesswork out of choosing among the many machine learning methods and their parameter settings prior to starting an analysis. Some of the early AutoML methods include Auto-Sklearn (Feurer et al. 2015), Auto-Weka (Thornton et al. 2013), and the Tree-Based Pipeline Optimization Tool, or TPOT (Olson et al. 2016a; Olson and Moore 2019). With each of these methods a computer algorithm searches for the optimal combination of machine learning algorithms and parameter settings to perform some prediction. Auto-Sklearn and Auto-Weka use Bayesian optimization and TPOT uses genetic programming. These methods have great potential and, if combined with the additional behaviors listed above, could be developed into full-fledged AI approaches to data analysis that would complement the work of human data scientists.

The goal of this chapter is to explore and review the role of AI in scientific discovery from data. Specifically, we will present AI as a useful tool for advancing a No-Boundary Thinking (NBT) approach to bioinformatics and biomedical informatics. No-Boundary Thinking is an agnostic methodology for scientific discovery and education that accesses, integrates, and synthesizes data, information, and knowledge from all disciplines to define important problems, leading to innovative and significant questions that can subsequently be addressed by individuals or collaborative teams with diverse expertise (Huang et al. 2013, 2015). Given this definition, AI is uniquely poised to advance NBT as it has the potential to employ data science for discovery by using information and knowledge from multiple disciplines. We present three recent AI approaches to data analysis that each contribute to a foundation for an NBT research strategy by either incorporating expert knowledge, automating machine learning, or both. We end with a vision for fully automating the discovery process while embracing NBT.

2.2 EXPLORATORY MODELING FOR EXTRACTING RELATIONSHIPS USING GENETIC AND EVOLUTIONARY NAVIGATION TECHNIQUES (EMERGENT)

A limitation of many statistical and machine learning methods is that they use a fixed mathematical framework that is applied to the data to be modeled. This can be a big assumption, depending on the data to be analyzed and the modeling approach that is selected. For example, choosing logistic regression as the modeling approach might not yield good results if the patterns in the data are inherently nonlinear. We have previously developed methods such as symbolic discriminant analysis (SDA) that can discover the functional form of the model using a set of mathematical functions that can be combined in different ways (Moore et al. 2002, 2007). This allows flexibility in the type of model that is to be applied to any given dataset. The goal of the EMERGENT algorithm is to take this several steps forward by allowing it to not only discover the functional form of the model, but to also learn how to generate good models (Moore et al. 2008, 2009). That is, the EMERGENT algorithm was designed to be self-adaptive in how it generates models. We briefly review the essence of the algorithm and then summarize its application to identifying genetic risk factors for glaucoma (Moore et al. 2015). We then tie this back to the NBT strategy.

The first component of EMERGENT is the models or solutions. These are initially generated randomly from a set of mathematical functions and variables or features. Solutions are represented as stack-based lists for ease of manipulation in the computer. Each solution forms a symbolic discriminant function that takes the features as inputs and return a discriminant score that is used to differentiate cases from controls (i.e. a discrete endpoint). The quality of a solution is measured using is classification accuracy. The solutions live on a grid and compete with others for survival to the next generation through a type of multi-objective fitness evaluation called Pareto optimization that considers multiple measures of model quality, including accuracy and complexity. In this way, better solutions can survive and worse solutions can be discarded.

The second component is the solution operator level. The goal of this component is to generate variability in the solutions so that they can be improved over time. Each solution operator modifies a subset of solutions on the grid through several different functions that can add, delete, or modify features or mathematical functions in the solutions. A key component of the solution operator is that it can add or delete features in the models according to expert knowledge provided by the user. This allows the system to modify solutions using multiple different types of knowledge that could include sources such as protein–protein interactions (Pattin et al. 2011) or even prior machine learning (Moore et al. 2014) or data visualization (Moore et al. 2011) results. The system rewards those solution operators that generate better solutions over time.

The third component is the mutation operator that can modify the solution operators. Each mutation operator can act on a subset of solution operators and get rewarded for survival based on the improvement in performance of those operators over time. There is an overall mutation probability parameter as well.

The EMERGENT system is thus designed to be self-adaptive in its solution discovery process and has the ability take advantage of expert knowledge provided to the system. This is a type of AI because the models are not prespecified but discovered based on experience and system feedback. This system has been applied to the genetic analysis of several diseases, including primary open-ended glaucoma, or POAG (Moore et al. 2015). This study included more than 2200 subjects with and without POAG and included more than 485,000 single-nucleotide polymorphisms (SNPs) measured across the human genome. The goal of the EMERGENT analysis was to identify models that captured nonadditive effects of combinations of SNPs on risk of POAG. Here, entropy-based methods were used to measure pairwise SNP–SNP interactions, and these were provided to EMERGENT both as expert knowledge and as a measure of model quality in the Pareto optimization. Thus, the system was rewarded for generating models with SNP–SNP interactions. The algorithm was run 1024 times and the results aggregated to arrive at a final best model. The model consisted of six SNPs all with evidence of pairwise interactions. One of these is a known risk factor for POAG while several others had been associated with eye phenotypes in experimental studies. Several were novel discoveries that had not previously been implicated in POAG or other eye diseases. Interestingly, all the genetic factors had known biological connections to the vascular endothelial growth factor (VEGF) gene that is a known drug target for glaucoma.

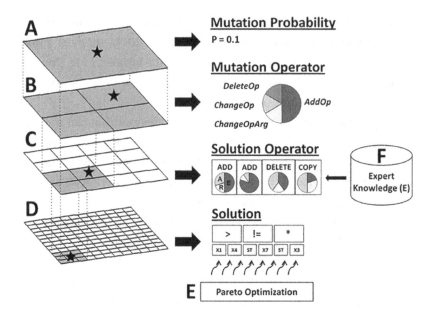

FIGURE 2.1 Overview of the EMERGENT system showing the overall mutation probability (A), the mutation operator (B), the solution operator (C), the solution grid (D), multi-objective Pareto evaluation of solutions (E), and expert knowledge sources (F) used by the solution operators. Note that the solutions are represented as stack (ST)-based lists.

The results of this study provided several insights. First, an AI could discover and exploit expert knowledge in the form of pre-computed SNP–SNP interactions to develop models that capture non-linear effects on risk of POAG. Second, the AI could identify best models which consisted of both known and novel genetic factors related to a known drug target. Third, the AI could self-adapt its model discovery strategy. This work is important for developing NBT strategies because it demonstrates use of expert knowledge by AI as it learns how to generate good models. Central to the NBT approach is the idea that knowledge from multiple disciplines is used to formulate and answer a scientific question. The EMERGENT algorithm is a step in this direction. Figure 2.1 provides a summary of the EMERGENT system.

2.3   TREE-BASED PIPELINE OPTIMIZATION TOOL

There are dozens of supervised machine learning algorithms available in the popular open-source scikit-learning or sklearn library (Pedregosa et al. 2011). Further, each of these algorithms has many possible parameter settings. It is hard to know beforehand which algorithm and parameter setting to pick, given they each look at the data in a different way (Olson et al. 2017, 2018a). Further, it can be computationally expensive to carry out a grid-search to manually explore many of the possible combinations. This is especially true if feature selection and feature transformation algorithms, for example, are added to the list. This has motivated the development of AutoML approaches for optimizing the selection and tuning of machine learning algorithms. Two of the first AutoML methods include Auto-Sklearn (Feurer et al. 2015) and Auto-Weka (Thornton et al. 2013), which each use Bayesian optimization to automate this process. We present here our own TPOT algorithm (Olson et al. 2016a, 2016b; Olson and Moore 2019) and provide some examples of how expert knowledge has been used to guide the AutoML process.

The goal of TPOT is to identify a pipeline of machine learning algorithms that provides an optimal prediction of a discrete or continuous outcome. Here, we utilize the machine learning algorithms available in the open-source and Python-based scikit-learn library (Pedregosa et al. 2011). The algorithms are the ingredients and the pipeline is the recipe. With TPOT, we represent pipelines as binary expression trees with the root node producing the output. This is a convenient data structure given trees are easily traversed with modern programming languages such as Python. Figure 2.2 provides a hypothetical pipeline that could be discovered by TPOT.

Central to TPOT is the exploration and evaluation of different pipelines. We use a stochastic optimization approach called genetic programming (GP) that starts by generating many pipelines (e.g. 1000 or more) randomly and then evaluating each for their ability to predict the endpoint. As with EMERGENT, TPOT performs a type

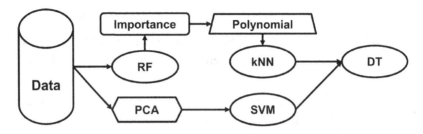

FIGURE 2.2 An example machine learning pipeline that could be constructed by TPOT. Here, the data are analyzed by two different branches of the pipeline. On the top, the data are analyzed by a random forest (RF) followed by feature selection using feature importance score and then a polynomial transformation. These new features are then analyzed by a k-nearest neighbor (kNN) algorithm, with those results fed to a decision tree (DT). On the bottom, the data are analyzed by principal component analysis (PCA) as a feature engineering step. The PCs are then fed to a support vector machine (SVM). The output of the SVM is then combined with the output of the kNN in a final DT analysis that performs the final classification.

of multi-objective fitness evaluation called Pareto optimization that considers multiple measures of model quality, including accuracy and complexity. The goal is to maximize the accuracy of the classifier (for discrete endpoints) and minimize the complexity of the pipelines to reduce overfitting and improve interpretation. With GP, the Pareto-optimal models are selected and then varied by randomly mutating pipeline components and by swapping pieces of pipelines using a recombination operator. In this way, new pipelines are created and evaluated. This process continues until a final best pipeline is selected. The TPOT approach has been evaluated (Olson et al. 2016a, 2016b; Olson and Moore 2019) and extended to big data problems. We summarize those studies next. TPOT has been recently reviewed in more detail (Olson and Moore 2019).

The goal of our first example was to adapt TPOT to the analysis of the type of genome-wide association study (GWAS) data analyzed in the EMERGENT example (Sohn et al. 2017). Most GWAS datasets consist of hundreds of thousands or millions of genetic variants or features. Thus, the data are much wider than they are deep with instances or subjects. We made two additions to TPOT for this large-scale genetic analysis. First, we added a feature engineering algorithm called multifactor dimensionality reduction (MDR) that is designed to detect SNP–SNP interactions (Ritchie et al. 2001). Second, we added a special expert-knowledge feature (EKF) selector operator in TPOT to perform feature selection using expert knowledge. With the EKF operator, the user can provide a ranking of the SNPs that is determined by some source of expert knowledge. The EKF then can select 1–5 top features, with the number selected being a tunable parameter. We demonstrated using simulated data that this version of TPOT was able to identify the correct two interacting SNPs with a higher success rate than TPOT without these operators. We then applied it to real GWAS data from the Cancer Genetic Markers of Susceptibility (CGEMS) study that focused on identifying genetic risk factor for aggressive forms of prostate cancer. More than 500,000 SNPs from this study were first filtered to just those in several relevant biochemical pathways. We then analyzed this set of SNPs using several different ReliefF algorithms (Moore 2015; Urbanowicz et al. 2018), including spatially uniform ReliefFF or SURF (Greene et al. 2009) to generate expert knowledge about SNP–SNP interactions. We demonstrated that TPOT with MDR and the EKF operator was able to identify better models of SNP–SNP interactions than other machine learning algorithms (Sohn et al. 2017).

The goal of the second example was to scale TPOT to genome-wide data with thousands or millions of genomic features (Le et al. 2020). For this task, we made two new modifications to TPOT. First, we added some new constraints to allow a template of a simple pipeline to be specified. For example, it might be of interest to only explore pipelines consisting of a feature selector followed by a feature transformer followed by one machine learning algorithm. These simpler constrained pipelines will be much more computationally efficient for the analysis of big data. Second, we added a new feature set selector (FSS). Here, a large GWAS dataset is first decomposed into many smaller subsets using expert knowledge such as biochemical pathways. Each smaller dataset is given an index number that serves as a parameter in the FSS operator in TPOT. This allows TPOT to select a small dataset to be included in the pipeline, thus improving the computational efficiency. We demonstrated using both simulated and real RNA sequencing data from a study of depression (Le et al. 2018) that this new approach could identify meaningful patterns in a computationally efficient manner (Le et al. 2020). The TPOT source code is freely available from Github (https://github.com/EpistasisLab/tpot).

The contribution of TPOT to the NBT approach is that it automates both the machine learning and the use of expert knowledge to assist with the analysis of biomedical big data. This will facilitate an analytical approach that uses information and knowledge from multiple disciplines to identify a predictive model of a specific disease or health outcome. For example, it might be of interest to predict depression from genomics data. However, genomic associations from multiple different diseases unrelated to depression could be used as expert knowledge for TPOT or EMERGENT, allowing these algorithms to exploit more general disease mechanisms.

2.4 ACCESSIBLE ARTIFICIAL INTELLIGENCE FOR DATA SCIENCE USING PENNAI

The EMERGENT and TPOT algorithms described above are designed to take the guesswork and assumptions about best models out of the machine learning process. This final example focuses on making AutoML easy and accessible so that this technology can reach anyone that wants to use it regardless of experience or expertise. We briefly describe here the PennAI method and software that is designed to bring AutoML to the masses (Olson et al. 2018b). PennAI represents an important step toward AI-driven data science by incorporating some of the features common to how humans approach and execute data analysis. Importantly, PennAI has a memory of all completed machine learning analyses and can learn from that experience to recommend new analyses to run. We briefly describe the different components of the PennAI system.

The first component of PennAI is the machine learning library that represents the various options for analysis. As with TPOT, we use the popular open-source scikit-learning or sklearn library that is programmed in Python (Pedregosa et al. 2011). The second component is a controller that can select a dataset, select a machine learning algorithm, run it on a server, and format the results for the system memory. We use the open-source Future Gadget Lab software (https:// kaixhin.github.io/FGLab) as our controller. The third component is a memory that is required for the system to learn over time what machine learning algorithm should be used for a dataset. This is key for the AutoML component of PennAI. Here, we use MongoDB as an open-source and freely available document store. Each machine learning result, the details of the data that were analyzed, and the details of the algorithm and parameter settings are encoded in JSON format and deposited into the MongoDB document store. The document store database makes it easy to deposit and retrieve machine learning results for use by the AI. The fourth component is the AI algorithm that analyzes the machine learning results in the database to make a recommendation on which machine learning algorithm should be run on the data next. We use a singular value decomposition (SVD) algorithm as part of the recommender system. The AI also has access to meta-features that describe the distribution of the data, types of features, the sample size, and other features such as the correlation structure. This provides the context with which certain machine learning algorithms might work better. The final component is a user-friendly interface that makes PennAI accessible to all. This was developed in JavaScript and is available through a web browser. Figure 2.3 gives a visual overview of these components. The system is built using Docker containers and is installed as an isolated software package on the user's secure server.

We have benchmarked PennAI (La Cava et al. 2020) using the Penn Machine Learning Benchmark (PMLB) resource (Olson et al.

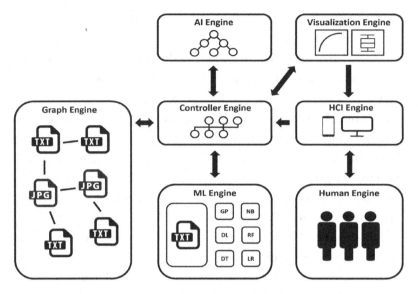

FIGURE 2.3   An overview of the PennAI system showing the machine (ML) component, the controller, the graph database, the AI component, and the visualization and human–computer interaction (HCI) components.

2017) that includes more than 150 real and simulated datasets that have been carefully characterized. Here, we showed that PennAI achieves near-peak performance after experiencing 50–100 datasets. This demonstrated that PennAI does in fact learn over time and improves its ability to recommend a good machine learning algorithm for a given dataset. This study compared the SVD recommended to several other approaches and showed that it was the top performer. We also applied it to a real dataset derived from the electronic health records of patients admitted to the intensive care unit and were able to show PennAI performed as well as a deep learning neural network for diagnosing septic shock, and was more computationally efficient (La Cava et al. 2020).

We live in a big data world, and there is no reason why everyone who wants to use computational approaches as part of their analytics strategy shouldn't have access to user-friendly machine learning. PennAI is a step in this direction and could serve as a platform for taking an NBT approach to data science. Extensions to PennAI could include the ability to use expert knowledge, as was implemented in several different ways as part of EMERGENT and TPOT. The goal would be to provide extensive sources of expert knowledge and to let PennAI learn which sources and which specific pieces of information and knowledge are most useful for solving a problem with machine learning. This would advance the NBT approach by harnessing knowledge generated by multiple different disciplines. We provide some additional ideas in the next section.

2.5 AN ARTIFICIAL INTELLIGENCE APPROACH TO NO-BOUNDARY THINKING

As we discussed above, NBT is an agnostic methodology for scientific discovery and education that accesses, integrates, and synthesizes data, information, and knowledge from all disciplines to define important problems, leading to innovative and significant questions that can subsequently be addressed by individuals or collaborative teams with diverse expertise (Huang et al. 2013, 2015). We think that AI can play an important role in developing an NBT approach to defining and solving problems in the biomedical sciences. In fact, we think AI will be required to implement an NBT strategy given the breadth and size of the biomedical knowledge base that would need to be accessed and synthesized. We discuss below a few examples of knowledge bases which could be incorporated into an AI strategy and then end with some thoughts about what an AI-based NBT platform might look like.

Perhaps the most useful source of expert knowledge that an AI could draw on is the scientific literature. A limitation of many biomedical research studies is that they only draw on literature from their own discipline. A cardiology study, for example, might not take advantage of the knowledge generated from nephrology. This is partly out of convenience since the biomedical literature grows at a rate that is not amenable to human consumption. There are just simply not enough hours in the day to read, digest, and synthesize findings from all biomedical areas, or even those in a single discipline. The first step is to develop algorithms that can retrieve the right scientific papers from the literature (Hersh 2009). The second step is to develop the natural language processing and text-mining algorithms that are needed to extract meaning (i.e. knowledge) from published papers that can be turned into structured information that can be used by an AI. This would facilitate automated hypothesis generation from the biomedical literature (Sybrandt et al. 2017) that would help realize an NBT approach. Structure data from literature could also serve as a source of expert knowledge to help guide machine learning methods such as TPOT and PennAI.

In addition to the scientific literature, there are many knowledge sources that might be useful for AI-driven data science. For example, the Hetionet database is a knowledge network derived from 29 various databases that include information about genes, drugs, diseases, etc. (Himmelstein and Baranzini 2015; Himmelstein et al. 2017). In this network, a gene and a disease are nodes with an edge connecting them if there is evidence for a biological relationship between them. Evidence leading to edges in the network can come from the literature or from biological experiments. All of this information is managed in an open-access Neo4J graph database. This resource can be used to provide biological interpretation of machine learning results or can be used as expert knowledge, as was done in the EMERGENT and TPOT projects. It can also be used to generate new hypotheses as part of an NBT approach. For example, a network of genes, drugs, and disease might reveal a cardiovascular drug that could be repurposed for cancer. This generates a hypothesis that could then be tested using big data or experimentally.

The key question is: Where are we headed with all these new technologies? How can knowledge bases and automated machine learning be combined to provide an AI approach to NBT? Figure 2.4 provides a vision for a connected set of resources and automated methodologies that does not yet exist, but could within the next 10 years. We hypothesize that a fully automated system like this could greatly accelerate scientific discovery and facilitate the NBT approach

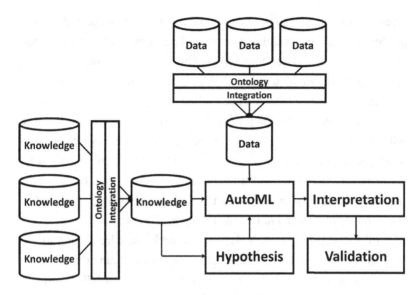

FIGURE 2.4 A vision for an automated approach to data science that embraces the NBT philosophy. Here, data and knowledge are integrated using ontologies and then fed to an AutoML algorithm that can be focused on a specific hypothesis derived from the knowledge base. AutoML results are interpreted and then validated. A goal would be to automate this entire process to speed up scientific discovery.

through articulating and answering new scientific questions that reach beyond any one discipline.

At the top of Figure 2.4 are several sources of data that can be integrated with the help of a biomedical ontology and other data standards and terminology. This might also represent a targeted subset of data, depending on the scientific question. To the right of the figure are different sources of knowledge that can also be integrated as in the Hetionet example from earlier. This integrated knowledge source could then be used in a directed or automated fashion to produce a hypothesis that would be addressed by an AutoML method that would combine the right data and the right knowledge to test the hypothesis. A final model would then be statistically and biologically interpreted and ultimately validated. The interpretation and validation could also be automated. Full automation of this computational discovery pipeline would allow the user to provide clean data and knowledge and receive an interpreted and validated model. The user could then decide whether the model is actionable. This would advance the NBT approach by allowing an automated system to be driven by a diversity of knowledge sources that are not limited to any one scientific domain. The hope is that this would produce results not anticipated by working within one domain. Although ambitious, this vision seems within reach given technology that is available today. There will no doubt be challenges to designing and implementing systems like this. Further, the results will only be as good as the data and knowledge provided to them. Inherent bias and noise will always be a problem to be addressed. Regardless, progress in methods for data and knowledge integration as well as AutoML provide a strong foundation for realizing this vision.

In summary, we have articulated a vision for an NBT approach to asking and answering biomedical research questions. The key to NBT is to move beyond the disciplinary silos we all feel comfortable in to articulate more impactful questions. We hypothesize that recent developments in AI and AutoML open the door to NBT approaches by making it easier to design analytic strategies that take advantage of data and knowledge across multiple disciplines. We have articulated a vision for an A–Z fully automated approach to discovery from data that embraces the NBT goal. This vision is within reach given current developments in AI and AutoML. It is our hope that this vision can be transformed into a research agenda in the coming years.

ACKNOWLEDGMENTS

This work was supported by National Institute of Health grants LM010098 and AI11679.

REFERENCES

Ching T, Himmelstein DS, Beaulieu-Jones BK, et al., 2018. Opportunities and obstacles for deep learning in biology and medicine. *J R Soc Interface*, 15. https://doi.org/10.1098/rsif.2017.0387

Crevier D, 1993. *AI: The Tumultuous History of the Search for Artificial Intelligence*. New York: Basic Books.

Esteva A, Robicquet A, Ramsundar B, et al., 2019. A guide to deep learning in healthcare. *Nature Medicine*, 25:24–29. https://doi.org/10.1038/s41591-018-0316-z

Ferrucci DA, 2012. Introduction to "This is Watson." *IBM Journal of Research and Development*, 56(1):1–15. https://doi.org/10.1147/JRD.2012.2184356

Feurer M, Klein A, Eggensperger K, et al., 2015. Efficient and robust automated machine learning. In: Cortes C, Lawrence ND, Lee DD, Sugiyama M, Garnett, R. (eds.), *Advances in Neural Information Processing Systems*. Red Hook, NY: Curran Associates, Inc., pp. 2962–2970.

Geng L, Hamilton HJ, 2006. Interestingness measures for data mining: a survey. *ACM Comput Surv*, 38. https://doi.org/10.1145/1132960.1132963

Greene CS, Penrod NM, Kiralis J, Moore JH, 2009. Spatially uniform reliefF (SURF) for computationally-efficient filtering of gene–gene interactions. *BioData Min*, 2:5. https://doi.org/10.1186/1756-0381-2-5

Hersh W, 2009. *Information Retrieval: A Health and Biomedical Perspective*, 3rd edn. New York: Springer-Verlag.

Himmelstein DS, Baranzini SE, 2015. Heterogeneous network edge prediction: a data integration approach to prioritize disease-associated genes. *PLoS Comput Biol*, 11: e1004259. https://doi.org/10.1371/journal.pcbi.1004259

Himmelstein DS, Lizee A, Hessler C, et al., 2017. Systematic integration of biomedical knowledge prioritizes drugs for repurposing. *Elife*, 6. https://doi.org/10.7554/eLife.26726

Hinton GE, Salakhutdinov RR, 2006. Reducing the dimensionality of data with neural networks. *Science*, 313:504–507. https://doi.org/10.1126/science.1127647

Huang X, Bruce B, Buchan A, et al., 2013. No-boundary thinking in bioinformatics research. *BioData Min*, 6:19. https://doi.org/10.1186/1756-0381-6-19

Huang X, Jennings SF, Bruce B, et al., 2015. Big data: a 21st century science Maginot Line? No-boundary thinking: shifting from the big data paradigm. *BioData Min*, 8:7. https://doi.org/10.1186/s13040-015-0037-5

Hutter F, Kotthoff L, Vanschoren J (eds.), 2019. *Automated Machine Learning: Methods, Systems, Challenges*. New York: Springer.

La Cava W, Williams H, Fu W, Moore JH, 2020. Evaluating recommender systems for AI-driven data science. *Bioinformatics*. https://doi.org/10.1093/bioinformatics/btaa698

Le TT, Savitz J, Suzuki H, et al., 2018. Identification and replication of RNA-Seq gene network modules associated with depression severity. *Transl Psychiatry*, 8:180. https://doi.org/10.1038/s41398-018-0234-3

Le TT, Fu W, Moore JH, 2020. Scaling tree-based automated machine learning to biomedical big data with a feature set selector. *Bioinformatics*, 36:250–256. https://doi.org/10.1093/bioinformatics/btz470

Moore JH, 2015. Epistasis analysis using ReliefF. *Methods Mol Biol*, 1253:315–325. https://doi.org/10.1007/978-1-4939-2155-3_17

Moore JH, Parker JS, Olsen NJ, Aune TM, 2002. Symbolic discriminant analysis of microarray data in autoimmune disease. *Genet Epidemiol*, 23:57–69. https://doi.org/10.1002/gepi.1117

Moore JH, Barney N, Tsai C-T, et al., 2007. Symbolic modeling of epistasis. *Hum Hered*, 63:120–133. https://doi.org/10.1159/000099184

Moore JH, Andrews PC, Barney N, White BC, 2008. Development and evaluation of an open-ended computational evolution system for the genetic analysis of susceptibility to common human diseases. In: Marchiori E, Moore JH (eds.), *Evolutionary Computation, Machine Learning and Data Mining in Bioinformatics*. Berlin: Springer, pp. 129–140. https://doi.org/10.1007/978-3-540-78757-0_12

Moore JH, Greene CS, Andrews PC, White BC, 2009. Does complexity matter? Artificial evolution, computational evolution and the genetic analysis of epistasis in common human diseases. In: *Genetic Programming Theory and Practice VI, Genetic and Evolutionary Computation*. Boston, MA: Springer, pp. 1–19. https://doi.org/10.1007/978-0-387-87623-8_9

Moore JH, Hill DP, Fisher JM, Lavender N, Kidd LC, 2011. Human–computer interaction in a computational evolution system for the genetic analysis of cancer. In: Riolo R, Vladislavleva E, Moore JH (eds.), *Genetic Programming Theory and Practice IX, Genetic and Evolutionary Computation*. New York: Springer, pp. 153–171. https://doi.org/10.1007/978-1-4614-1770-5_9

Moore JH, Hill DP, Saykin A, Shen L, 2014. Exploring interestingness in a computational evolution system for the genome-wide genetic analysis of Alzheimer's disease. In Riolo R, Moore JH, Kotanchek M (eds.), *Genetic Programming Theory and Practice XI, Genetic and Evolutionary Computation*. New York: Springer, pp. 31–45. https://doi.org/10.1007/978-1-4939-0375-7_2

Moore JH, Greene CS, Hill DP, 2015. Identification of novel genetic models of glaucoma using the "EMERGENT" genetic programming-based artificial intelligence system. In Riolo R, Worzel WP, Kotanchek M (eds.), *Genetic Programming Theory and Practice XII, Genetic and Evolutionary Computation*. Cham: Springer, pp. 17–35. https://doi.org/10.1007/978-3-319-16030-6_2

Olson RS, Moore JH, 2019. TPOT: a tree-based pipeline optimization tool for automating machine learning. In: Hutter F, Kotthoff L, Vanschoren J (eds.),

Automated Machine Learning: Methods, Systems, Challenges. Cham: Springer, pp. 151–160. https://doi.org/10.1007/978-3-030-05318-5_8

Olson RS, Bartley N, Urbanowicz RJ, Moore JH, 2016a. Evaluation of a tree-based pipeline optimization tool for automating data science. In: *Proceedings of the Genetic and Evolutionary Computation Conference 2016, GECCO'16.* New York: ACM, pp. 485–492. https://doi.org/10.1145/2908812.2908918

Olson RS, Urbanowicz RJ, Andrews PC, et al., 2016b. Automating biomedical data science through tree-based pipeline optimization. In: Squillero G, Burelli P (eds.), *Applications of Evolutionary Computation.* Cham: Springer, pp. 123–137. https://doi.org/10.1007/978-3-319-31204-0_9

Olson RS, La Cava W, Orzechowski P, Urbanowicz RJ, Moore JH, 2017. PMLB: a large benchmark suite for machine learning evaluation and comparison. *BioData Min,* 10:36. https://doi.org/10.1186/s13040-017-0154-4

Olson RS, La Cava W, Mustahsan Z, Varik A, Moore JH, 2018a. Data-driven advice for applying machine learning to bioinformatics problems. *Pac Symp Biocomput,* 23:192–203.

Olson RS, Sipper M, La Cava W, et al., 2018b. A system for accessible artificial intelligence. In: Banzhaf W, Olson RS, Tozier W, Riolo R (eds.), *Genetic Programming Theory and Practice XV, Genetic and Evolutionary Computation.* New York: Springer, pp. 121–134.

Pattin KA, Payne JL, Hill DP, et al., 2011. Exploiting expert knowledge of protein-protein interactions in a computational evolution system for detecting epistasis. In Riolo R, McConaghy T, Vladislavleva E (eds.), *Genetic Programming Theory and Practice VIII, Genetic and Evolutionary Computation.* New York: Springer, pp. 195–210. https://doi.org/10.1007/978-1-4419-7747-2_12

Pedregosa F, Varoquaux G, Gramfort A, et al., 2011. Scikit-learn: machine learning in Python. *J Mach Learn Res,* 12:2825–2830.

Ritchie MD, Hahn LW, Roodi N, et al., 2001. Multifactor-dimensionality reduction reveals high-order interactions among estrogen-metabolism genes in sporadic breast cancer. *Am J Hum Genet,* 69:138–147. https://doi.org/10.1086/321276

Sohn A, Olson RS, Moore JH, 2017. Toward the automated analysis of complex diseases in genome-wide association studies using genetic programming. In: *Proceedings of the Genetic and Evolutionary Computation Conference, GECCO'17.* New York: ACM, pp. 489–496. https://doi.org/10.1145/3071178.3071212

Strickland E, 2019. IBM Watson, heal thyself: how IBM overpromised and under-delivered on AI health care. *IEEE Spectrum,* 56:24–31. https://doi.org/10.1109/MSPEC.2019.8678513

Sybrandt J, Shtutman M, Safro I, 2017. MOLIERE: automatic biomedical hypothesis generation system. In: *Proceedings of the 23rd ACM SIGKDD International*

Conference on Knowledge Discovery and Data Mining, KDD'17. New York: ACM, pp. 1633–1642. https://doi.org/10.1145/3097983.3098057

Thornton C, Hutter F, Hoos HH, Leyton-Brown K, 2013. Auto-WEKA: combined selection and hyperparameter optimization of classification algorithms. In: *Proceedings of the 19th ACM SIGKDD International Conference on Knowledge Discovery and Data Mining, KDD'13.* New York: ACM, pp. 847–855. https://doi .org/10.1145/2487575.2487629

Topol EJ, 2019. High-performance medicine: the convergence of human and artificial intelligence. *Nat Med,* 25:44–56. https://doi.org/10.1038/s41591-018-0300-7

Urbanowicz RJ, Meeker M, La Cava W, Olson RS, Moore JH, 2018. Relief-based feature selection: introduction and review. *J Biomed Informat,* 85:189–203. https://doi.org/10.1016/j.jbi.2018.07.014

3 No-Boundary Thinking in Undergraduate Bioinformatics Education

Yu Zhang, Clare Bates Congdon, and Matt Hibbs

3.1 INTRODUCTION

Bioinformatics is one of the fastest growing fields in the twenty-first century (Tymann et al. 2005). Over the last few decades, studies of biology have moved from low-throughput hands-on experiments to computational analyses of the increasingly complex tree of life. Scientists from biology, computer science, chemistry, mathematics, physics, and possibly other disciplines can use many computational programs to access DNA, RNA, and protein sequence data in order to simulate actual life and revolutionize almost everything from medicine to food science (Greengard 2014). This is to say, this new frontier of computational biology and bioinformatics is changing our world (Berger et al. 2016).

This change presents multiple challenges. The first challenge exists in *interdisciplinary collaboration*. The current interdisciplinary collaboration model is still bounded by individual disciplines and is far from seamless. Clearly, a working understanding of bioinformatics requires a synthesis of principles from biology and computer science as well as applied mathematics, quantum physics, and chemistry (Maloney et al. 2010). If we simply put a problem in a production pipeline, starting with the biologists who collect data from wet labs, passing through mathematicians, physicists, and chemists who work on quantitative models, and ending with the computer scientists who run the data and the model on computers, then we cultivate restrictive boundaries around each area of expertise. This way, each scientist does not know each other's work and so cannot have much insight to the whole problem. This problem-solving structure negatively affects even

25

the process of defining the questions to be addressed in the work, as each scientist will necessarily understand the work to be done within the frameworks of their own fields. No-Boundary Thinking (NBT) (Huang et al. 2013) is a better interdisciplinary framework that has all the experts start to work together at the beginning to create a boundary-free problem definition and solution.

The second challenge is *big data*. Over the past few decades, major advances in the field of molecular biology, coupled with advances in genomic technologies, have led to an explosive growth in the biological information generated by researchers. Big data is often characterized by the four Vs: volume, velocity, variety, and veracity. Volume refers to the quantity of data; velocity refers to its speed of creation; variety refers to heterogeneous sources and structures of the available data; and veracity refers to the "correctness" of the recorded data. Each of these factors plays a role in bioinformatics work. For example, multiple sources of data bring up the need to clean, match, and link the data into a uniform database so we can easily search and query the data. More importantly, the lack of knowledge of the underlying empirical micro-processes that lead to the emergence of the network characteristics of biological data creates a difficulty in knowing how best to design such a database (Nasser and Tariq 2015). Biology has too often developed and embraced standard analytical methods and representations that do not attempt to understand complexity, but instead remove uniqueness of a given trait where it needs to remain (Williams and Moore 2013). But what we actually care about is the useful data and how to extract them from the haystack of big data. The big data challenge again emphasizes the NBT approach because the complexity of the data needs complete interpretations from different experts and a closer collaboration from the beginning of the research.

The third challenge is *human infrastructure*. We need to educate the next generation of scientists as early as possible, including at the undergraduate level, to solve the interdisciplinary and complex biology problems with computational resources. Most current undergraduate educational approaches still use traditional approaches that have not changed for 50 years. Students are established in a discipline from the beginning. They may be trained by an interdisciplinary, multidisciplinary, or transdisciplinary (Choi and Pak 2006) program for a few courses, but all of them still have boundaries between disciplines. No-Boundary Thinking is a new process to free students from their own discipline. For example, a biology major can understand how computational models of biological phenomena are created, and a computer science major can understand the known biological mechanisms to be explored via computational work. To this end, our next generation of scientists won't be just workers sitting in a production pipeline only focusing on the teachings of one traditional discipline, but instead they will be broadly educated thinkers who won't know everything in all fields, but can transcend disciplinary training and approach to a problem's definition in its broadest terms by knowing how to ask good questions, and how to bring the right people and the right disciplines to fill in the blanks. Importantly, the NBT scientists of the future will be able to converse with others who have different expertise in order to work together to define and solve problems.

To address these challenges, the National Research Council has responded with the New Biology initiative (Labov et al. 2009). The essence of New Biology is two types of integrations: the integration of all sub-disciplines of biology and the integration into biology of other disciplines such as computer science, physics, chemistry, engineering, and mathematics. Other institutions or universities have also reformed their own curricula and research to create capacities to tackle a broad range of scientific problems (e.g. Chapman et al. 2004; Howard et al. 2007; Miskowski et al. 2007; Ditty et al. 2010; Irwin and Shoichet 2016; Beheshti et al. 2017). But most of these focus on adding computational tools and programs to biology, but do not add biological aspects to a computer science perspective. This chapter fills this gap by introducing teaching and research practices incorporating biological questions to the education of computer scientists.

3.2   BIOINFORMATICS AND UNDERGRADUATE
CURRICULA: A REVIEW

In this section we review a few representative bioinformatics under-graduate curricula and projects that integrate research into teaching and integrate other disciplines into biology. Undergraduate research has been widely believed to be an effective way to retain students in STEM (science, technology, engineering, and mathematics) fields and encourage them to pursue advanced degrees and careers in these fields (Russell et al. 2007). Traditionally, summer research experiences or an independent study course used to be the only format of bioinformatics research in most undergraduate programs. Now there are more and more successful integrations of research into undergraduate curricula.

3.2.1   Collaborative Networks

The Genomics Education Partnership (GEP) is a nationwide faculty col-laborative network on genomics involving both research-intensive uni-versities and primarily undergraduate institutions (Shaffer et al. 2010; Lopatto et al. 2014). This project aims to provide undergraduates with a research experience in genomics through a scheduled course (a classroom-based undergraduate research experience, or CURE). The course could be a short module of a first genetics course, a semester-long core laboratory course, or an independent study research course. Participating faculty receive training on research tools such as R statistical language, as well as curriculum developments. The central GEP support system maintains up-to-date access to the research tools and a public website, and shares all curricula, datasets, and student assess-ments. The network grows by hosting summer training workshops for new faculty. The network presently has over 100 universities and most of them are primarily undergraduate institutions. Effectively, this collabora-tive effort saves the cost of equipment, supplies, and support for trained mentors, especially for small schools with limited financial budgets.

Integrated Microbial Genomes Annotation Collaboration Toolkit (IMG-ACT) is another collaborative project funded by the Department of Energy – Joint Genome Institute. IMG-ACT and GEP (which is jointly funded by the National Institutes of Health and the Howard Hughes Medical Institute) are similar in many ways. For example, they both (1) run as a means to innovate and update undergraduate biology education and faculty development; (2) focus on real data and tools that reflect the cutting edge research in bioinformatics; (3) integrate research into a regular course or an independent study course; (4) maintain an open-access infrastructure that share data, courses, and research tools and projects; and (5) involve research-intensive and predominantly under-graduate institutions. The major difference between IMG-ACT and GEP is the research problems they pick up. IMG-ACT's project is about microbial genome analysis and GEP is about a particular portion of the dot chromosome (Muller F element) in *Drosophila*. In spite of this minor difference, the two projects are both successful implementations from research to undergraduate curricula.

3.2.2 Bioinformatics Curricula in Liberal Arts Institutions

Both IMG-ACT and GEP projects are large collaborative networks connecting research-intensive universities and primarily undergradu-ate institutions. However, in some small liberal arts colleges without such connections, collaborations in the sciences can be difficult to implement in undergraduate classrooms, especially in a relatively new field such as genomics. This is not to say that collaborations can't be done in small colleges. Actually, most experiences gained from the large collaborative networks can be implemented in small colleges without any change. Moreover, the small size of the colleges may make the collaboration even easier in the sense that faculty members are more likely to interact with faculty members from other departments – for example, in campus governance. This may help communicate with each other across disciplinary boundaries. We summarize three practices that work for small liberal arts colleges.

1. **Compare and contrast programs.** Departments within the same college or different colleges in the same geographic area can frequently meet to

compare and contrast each other's disciplinary programs. In this way, faculty can exchange ideas on teaching and research and suggestions for homework and projects, and can share biology and algorithmic resources (Dyer and LeBlanc 2002).

2. **Linked courses.** Biology and computer science faculty can team up to develop courses that share common parts and teach them in each other's classes. For example, infusing genomics into both a biology course and an algorithm course and teaching it within each other's context; or infusing network theory into an introductory biology course and a data structure course so students know how to simulate genomics problems by regulatory networks (Banta et al. 2012).

3. Communication across the disciplinary boundaries. It is not easy to understand any discipline if it is not your field. This is truer in the sciences because of their uneven evolution (for example, biology is an "old" science, while computer science is a "new" science). But even in the old sciences such as biology there are new fields such as genomics, and in the new sciences such as computer science there are newer concepts such as big data and cloud computing. Therefore, communication across the disciplinary boundaries is very important and helpful for everybody. We need to learn the "right" way to ask the "right" question of the "right" person (Maloney et al. 2010).

3.2.3 Integrated Science

Integrated science is a revolutionary introductory science curriculum developed at Princeton University (Bialek and Botstein 2004). The program provides a series of courses taken in the freshman and sophomore years to all students who want to major in any of the core scientific disciplines. Students are exposed to fundamental concepts in biology, computer science, mathematics, and physics. These concepts are integrated in the same courses rather than being taught individually. For example, calculus is taught together with basic physics; thermodynamics and kinetic theory appear in different introductory courses to physics, chemistry, and biology; quantum theory to describe electrons, orbitals, and chemical bonding is taught in an introductory chemistry course, etc. Another feature of this program is to include biology in the quantitative and mathematical culture that

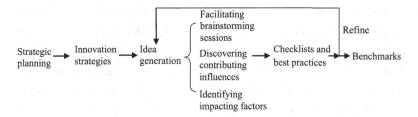

FIGURE 3.1   Steps in problem definition.

has come to define the physical sciences and engineering. For example, statistics in experimental design; pattern recognition in bioinformatics; models in evolution, ecology, and epidemiology. Actually, all courses promote quantitative thinking and encourage students to investigate connections between disciplines.

3.3   A NO-BOUNDARY THINKING APPROACH TO TEACHING THE NEXT GENERATION OF SCIENTISTS

No-Boundary Thinking encourages scientists to transcend their disciplinary training and approach a problem definition in its broadest terms (Huang et al. 2013). At the top level of almost any field, people are distinguished not by what they know but how they deal with the unknown. The NBT problem-defining approach includes a broad range of disciplinary methodologies. We present a high-level implementation to the concept of NBT in Figure 3.1.

Faculty mentors and students in each project will work closely together, starting from *strategic planning* on making decisions on the research direction, allocating the resources (such as the 10-week summer research time, labs and equipment), and defining the control mechanism for guiding the experiment. Among these strategies, the *innovation strategies* are the most important because they will decide the underlying science challenges that will advance the science as a whole. Next, faculty and students will *generate ideas* in order to identify the state-of-the-art research from each discipline and integrate them to produce new algorithms and mathematical models. Formal or informal sessions will be held to *facilitate brainstorming,* to *discover contributing influences* to the project, and to *identify factors* that impact potential solutions. To effectively manage the project, each group will maintain a *checklist and best practices* for both the group and the individuals. The brainstorming sessions will repeat to refine the checklist and practices. The ultimate outcome of this procedure is the *benchmark* that defines the settings and operations to assess algorithms and mathematical models.

Our approach is a top-down method (shown from left to right in Figure 3.1). Traditional bioinformatics research is bottom-up, or data-driven – that is, information processing and knowledge ordering start from data, including the form and scale of the data and missing data, followed by association studies to find links among the data. This process is repeated until some significant hypothesis is generated. This method is nice, but the problem is that it does not show whether these associations have meaning. Finding a signal is only the first step (Khoury and Ioannidis 2014). We can address this problem well because our approach is a top-down method. It is inspired by questions raised in the study of biology – from curing cancer to fighting poverty. These real-world challenges present a spectrum of needs for information and knowledge across multiple disciplines. Scientists need to collaborate at the beginning to define the problem together. This way they can minimize the noise in the large-scale data and minimize biases in selection, as well as deal with confounding variables and lack of generalizability of the data.

We take a broad approach to problem definition. Problem definition and problem solving are inseparable parts of problem resolution. But how much time you will give to each one? Albert Einstein answered this question: "If I were given one hour to save the planet, I would spend 59 minutes defining the problem and one minute resolving it."[1] These were wise words, but from what we have observed the current undergraduate education fosters problem-solving skills to build students' confidence in dealing with the unknown but

---

[1] https://hbr.org/2012/09/are-you-solving-the-right-problem.

seldom asks them whether they are solving the right problem. We emphasize having a well-defined problem, or a statement of the problem, before involving any problem-solving methods. The process of problem definition is also the process of information gathering. When we have an unknown problem, we want to ask ourselves these questions: What are the possible causes of the problem (causes can be broad, such as people, resources, environment, process, procedures)? Are these causes only related to my discipline? If not, what other disciplines do they involve? How can we find the people with the right backgrounds from other disciplines? How can we learn from each other to be innovative? How should we manage risk? To this end, a broad approach to problem definition can establish the right problem and ultimately minimize the time spent solving the problem.

Our approach focuses on communicating science. Bioinformatics research refers to a broad range of activities involving biology, computer science, physics, mathematics, chemistry, and more. Communicating science teaches scientists to discuss complicated scientific problems in a clear, concise manner with people outside their immediate field. The ability to convey complex concepts effectively can help scientists successfully engage in a variety of public and professional interactions, such as research collaborations, conducting media interviews, project management, writing grant proposals, discussing ideas with students, and speaking in public forums (Gross et al. 2007).

We encourage scientists with different backgrounds to define a "communication differences" table at the beginning of the research and to clarify any terminologies and concepts as early as possible. The communication table may be extended along with the development of the project until it is done. For example, BIOAGENTS is an agent-based approach to computational frameworks for both data analysis and management in bioinformatics (Merelli et al. 2006). Here, "agent" is a technical terminology different from its dictionary definition. A computer scientist can explain the agent terminology to non-computer scientists at three levels, from the general to the specific: (1) an agent is a high-level software abstraction or, simply, a software program; (2) an agent can be viewed as an autonomous object that can perform problem-solving actions such as managing the primary databases, performing sequence analyses using existing tools, and storing and presenting resulting information; and (3) there can be multiple agents, or a multi-agent system, that act in different roles and interact with the environment, such as databases, and share their database information with each other.

3.4   THREE NO-BOUNDARY THINKING TEACHING AND RESEARCH MODELS

3.4.1   An Interdisciplinary Undergraduate Seminar

One of the challenges in learning to do multidisciplinary work is to learn to talk to others whose core discipline is different from one's own. Thus, an ideal (if resource-intensive) approach to learning NBT thinking may be to offer a multidisciplinary course, open to students from different disciplines and co-taught by professors from different disciplines. In spring 2003 such a course was first offered at Colby College (a version of this course was later offered at the University of Southern Maine in spring 2011 and spring 2013). The senior-level course in bioinformatics was cross-listed in the Departments of Biology and Computer Science, and was taught without requiring students to have prerequisites from the other discipline. Both professors were present at all class meetings.

During the first several weeks of the course, students covered the first few chapters of an interdisciplinary undergraduate bioinformatics text (Krane and Raymer, 2003), with the goal of completing a quick and broad survey of core topics from both biology and computer science. Rather than lecturing to the students, students were required to use an online discussion board before class time, both to pose questions and to try to answer the questions of other students. The task and the challenge during this phase of the course is to try to get students to answer the questions pertaining to their discipline for students not from their discipline, and to avoid using jargon when doing so (or to explain their discipline's jargon as part of the answer). Questions and online discussions between class meetings were gathered on a single document by the professors as a guide for in-person classroom discussions. Again, in class there was an emphasis on students explaining things to students not from their own discipline. In both the online discussion and in the classroom professors stepped in to guide conversations that were going off track in correctness or focus, to remind students to avoid the jargon of their discipline, and to help clarify concepts that were difficult for students to explain to each other.

This phase of the class is enormously important to laying the groundwork for NBT education, perhaps especially for undergraduates, who have been working so hard to learn the jargon of their discipline and must now take a step or two back from that training to communicate effectively with others. During this phase the emphasis was not on "learning all the bioinformatics ideas," but learning to talk about those concepts and ideas that were most compelling to the students in the class while gaining an understanding of the breadth and depth of work that is called "bioinformatics." There are some core ideas from each host discipline that must be covered and understood in this phase; computer science students must learn the central dogma, for example, and biology students must learn what an algorithm is. Specific tasks such as alignments and phylogenetics were covered during this phase of the class and supplemented with hands-on labs to help make the concepts concrete.

As a mid-semester project, students from different disciplines paired up to pursue a particular bioinformatics topic in depth, using library and internet resources, and to bring this work back to the rest of the class in the form of a brief talk and an online paper or website. Topics covered at this point were driven largely by student interest and guided by the professor. Recent topics include sequence assembly, genome annotation, the ENCODE project, phylogenetic models of viral evolution, and modeling genetic epistasis.

The second half of the class is devoted to developing and pursuing modest original research projects in interdisciplinary pairs. After the first half of the semester, students have largely learned to communicate with each other, have sampled the basic ideas of the broad multidisciplinary field of bioinformatics, and have learned about some specific topic in more depth. Projects are chosen based on students interests; typically, several of the students have questions brought from previous coursework or research experiences that inform their project design; other projects are informed by research projects of the course professors or their colleagues. Example projects include investigating candidate regulatory regions for genes of interest, protein structure prediction for specific domains, bacteriophage sequence annotation, and investigations of gene expression data associated with cancers. Students present their work in multiple forms, including a talk to the class, a research poster, and a final paper.

3.4.2 An Undergraduate Seminar Model

In order to expose interested students to the broad area of bioinformatics, the Computer Science Department at Trinity University offers a seminar-style elective class in computational biology. This course has been taught in a manner to emphasize the goals of NBT, specifically by placing an emphasis on interdisciplinary communication and by approaching problems in a top-down fashion. The material for this course was presented through a selection of ~24 journal articles, ranging from bioinformatics "classics" (such as Needleman and Wunsch [1970]) through landmark developments (including the sequencing of the human genome [Venter et al. 2001]) to highly recent advances (such as metagenomics [Zhernakova et al. 2016]). Each 75-minute class meeting was structured around a single article being presented by a student for the first 45 minutes, followed by enriching material presented by the instructor and discussion by the class. The presenting student met with the instructor at least one week before their class presentation to discuss the article and to get feedback on their presentation. All students were expected to read the planned article before class, and a comprehension quiz was used to assess both the student presenter and the audience members. Specifically, a brief short-answer quiz was created by the instructor and given to all audience students after presentations, and a portion of the presenter's grade was determined by the average quiz scores and a portion of the audience members' grades were determined by their own quiz performances.

While this course is offered to upper division computer science majors, who should possess a strong baseline of computational and algorithmic knowledge, this population of students demonstrates a wide range of biological backgrounds, ranging from only high school biology courses to students that double major in computer science and biology. As such, the first two weeks of the course were spent reviewing the core basics of biology in order to standardize a baseline of knowledge, including the nature of DNA, the "central dogma," Mendelian genetics, protein folding, and more. This initial review period proved useful for the majority of students in the class by helping orient them to the language and topics that would be further investigated throughout the semester. Biology is notorious for its complex and sometimes confusing terminology, and as such through-out the class – but especially during the initial review period – students maintained and updated a collective "terminology database," which was a shared document where students could request and/or provide definitions for biological terms. For example, a student could request a definition for "transcription" and rather than simply relying on a textbook-level answer (e.g. "transcription is the process of creating mRNA molecules from a DNA template"), fellow students could provide a definition appropriate to our class (e.g. "transcription is analogous to compiling a program written in DNA into an intermediate byte-code language written in mRNA"). The resulting document was among the most highly used tools by the students to help them understand later course material and communicate more clearly between disciplines.

In addition to the seminar-style literature review, students worked individually or in teams of two to approach an open bioinformatics research question throughout the semester as a stand-in for their final exam. The goal of this project was not necessarily for the students to completely solve an open area of research, but to gain a better understanding of the bioinformatics research process, identify pitfalls, brainstorm solutions, and pursue an approach to evaluate its effectiveness. As such, "success" was not defined as achieving a new solution to a problem, but as seriously investigating, understanding, and exploring an open research problem. This is in line with the NBT focus on "top-down" problem definition, since at their current level of training these students are better served by focusing on defining concrete problems, narrowing the scope, and then zeroing in on potential solutions, rather than simply casting a wide net and hoping for the best. For many students in this course this was their first exposure to a true research question, where a textbook or the Internet could not be used to look up or verify an answer. This experience was uncomfortable for many students, but was also eye-opening for many, highlighting the boundaries of our current knowledge and how they could expand those boundaries, even as students.

In order to enable this "controlled" research experience, especially for new researchers, the instructor provided a selection of 4–6 appropriately scoped open research questions that could be directly adopted by students and teams, or could be used as inspiration to define their own research questions. Each student or team met individually with the instructor at least three times throughout the semester for formal evaluations and to solicit feedback and advice. In our experience, this level of scaffolding should be seen as a minimum, as many students would benefit from additional checkpoints and individual meetings, both formal and informal.

Student projects were assessed largely through two written documents and two oral presentations to the class. About halfway through the semester, each student or team created a two-page "specific aims" document, following an NIH-style grant proposal, outlining their problem of interest and describing their potential approach and possible results. They also presented their aims to the class in brief presentations, where they received feedback and advice on how to move forward. At the end of the semester, each student or team created a "progress report" NIH-style document, detailing the results of their efforts. These results were also presented to the class as a final capstone summarizing their work during the last week of the semester. In the spirit of NBT, this approach encouraged students to thoughtfully define their problems in terms that their peers could understand and to practice their written and verbal communication skills, often discussing a discipline that was not their primary field of study.

3.4.3   An Undergraduate Thesis Model

Research beginners need a foundation in several fields to be able to work on the most important problems confronting scientists today. Nowadays, bioinformatics is not only about managing and analyzing database systems, but also integrating two other branches: (1) computational biology, which focuses on the algorithm aspects and the complexity of the system (Kessler et al. 2015); and (2) systems biology, which attempts to understand the emerging behavior of biological systems as a whole (Palsson and Palsson 2016).

In the Department of Computer Science at Trinity University, undergraduate students closely work with faculty on research projects during the academic year. We focus on designing and developing quantitative and computational models to solve biology problems. The research is designed for students with a strong interest in multidisciplinary and systems-level approaches to understanding molecular, cellular, and organismal behavior. A strong emphasis is placed on using global genome-wide measurements, such as microarray gene expression, gene sequence and phenotype, to understand physiological and evolutionary processes.

We implement an undergraduate thesis model based on the timeline shown in Table 3.1.

The research generally starts in the spring semester of the junior year, when students register for a thesis reading course worth 1–3 credits. This course reviews a breadth of research topics, methods,

Table 3.1 *A timeline model for undergraduate thesis*

Timeline	Spring of junior year	Summer of junior year	Fall of senior year	Spring of senior year
Progress	Thesis reading	Research problem definition	Research proposal; problem solving and experiment	Thesis and defense

and tools. It covers not only algorithms, artificial intelligence, and machine learning from computer science, but also probability theory and statistics from math, fluid dynamics, waves, and quantum chemistry from both physics and chemistry, molecular biology and genetic regulation from biology, and more. With the increasingly interdisciplinary approach to scientific research, exposure to such a wide breadth of topics is extremely valuable to students.

In the summer of the junior year, students typically are supported by a REU (Research Experiences for Undergraduates) program and spend 10 weeks engaged in intensive training on defining their research statements. Every summer most of the STEM departments have their own REU programs through agencies such as NSF, HHMI, Welch Foundation, Mellon Foundation, Murchison Foundation, McNair Scholars, and Noyce Foundation. This is a great way for research collaborations to happen among faculty and students. For example, computer science students may come to a biology or chemistry laboratory to learn experimental and analytic techniques for acquisition of biology data, and the interpretation of such data in the context of a quantitative model. During the summer, faculty and students are connected not only by individually scheduled meetings and seminars, but by the university-wide Trinity Summer Research Symposium and other conferences.

The senior year embraces a lot of rigorous scientific writing from a proposal to a thesis – from composing an effective and motivating abstract to designing and presenting scientific findings.

Research collaborations continue throughout the whole year. The student's thesis committee normally is composed of faculty members from two or three disciplines.

This thesis model implements the main perspectives of NBT.

- It is a top-down model. At the beginning, the central role of quantitative problem definition and computational modeling is emphasized. Students learn related disciplines through quantitative problem definition and computational modeling – that is, students systematically think about and compute every perspective of the problem.
- It takes a broad approach to problem definition. This broad approach is achieved in three ways. (1) Students have total freedom to find the scientific problem that really interests them. The role of faculty is to teach them how to think. This will help students develop the questioning and probing minds of scientific researchers. (2) Students are taught to tackle a problem from multiple viewpoints, either within their own discipline or beyond. Finding a solution is not the whole point of the research; defining the right problem by using all possible resources and knowledge is more critical. (3) The research simulates the experiences of real researchers in their attempts to answer questions that have never been answered, and in so doing it prepares students for research careers in a whole range of scientific disciplines.
- It involves a lot of communication. Science-making is not a result of individual endeavor, but the product of communal effort. Through this process students will feel comfortable talking to researchers in other disciplines to discuss all types of science. Using a wide variety of tools and approaches, students will know how to ask the right people the right questions.

3.5 CONCLUSION

The NBT approach includes a broad range of institutional and disciplinary methodologies. In this chapter we present three models, all in undergraduate teaching and research. Our method focuses on continual communication with other disciplines. We made an extensive effort to learn the language of other fields and to teach others the language of our own. Through the process we appreciated the value of both methods as well as understood the limitations of both. We have all recognized that this effort must happen before successful collaboration can occur. Second, we tried not to guess the result early, but left the question open until we believed we had enough data to form a sensible hypothesis. Third, we considered human resources, our young generation of researchers, as one of the major factors in the NBT infrastructure. Our training centered on nurturing undergraduate students who in the future will push for scientific breakthroughs in computational biology and chemistry with no-boundary problem definition and problem solving.

REFERENCES

Banta LM, Crespi EJ, Nehm RH, et al., 2012. Integrating genomics research throughout the undergraduate curriculum: a collection of inquiry-based genomics lab modules. *CBE Life Sci Educ*, 11:203–208.

Beheshti I, Demirel H, Matsuda H, 2017. Classification of Alzheimer's disease and prediction of mild cognitive impairment-to-Alzheimer's conversion from structural magnetic resource imaging using feature ranking and a genetic algorithm. *Comput Biol Med*, 83(1):109–119.

Berger B, Daniels NM, Yu YW, 2016. Computational biology in the 21st century: scaling with compressive algorithms. *Commun ACM*, 59(8):72–80.

Bialek W, Botstein D, 2004. Introductory science and mathematics education for 21st-century biologists. *Science*, 303(5659):788–790.

Chapman BS, Christmann JL, Thatcher EF, 2004. Bioinformatics for undergraduates: steps toward a quantitative bioscience curriculum. *Biochem Mol Biol Educ*, 34:180–186.

Choi B, Pak A, 2006. Multidisciplinarity, interdisciplinarity and transdisciplinarity in health research, services, education and policy: definitions objectives and evidence of effectiveness. *Clin Invest Med*, 29(6):351–364.

Ditty JL, Kvaal CA, Goodner B, et al., 2010. Incorporating genomics and bioinformatics across the life sciences curriculum. *PLoS Biol*, 8(8):1–4.

Dyer BD, LeBlanc M, 2002. Meeting report: incorporating genomics research into undergraduate curricula. *Cell Biol Educ*, 1:101–104.

Greengard S, 2014. How computers are changing biology. *Commun ACM*, 57 (5):21–23.

Gross AG, Harmon JE, Reidy M, 2007. *Communicating Science: The Scientific Article from the 17th Century to the Present*. Oxford: Oxford University Press.

Howard DR, Miskowski JA, Grunwald SK, Abler ML, 2007. Assessment of a bioinformatics across life science curricula initiative. *Biochem Mol Biol Educ*, 35:16–23.

Huang X, Bruce B, Buchan A, et al., 2013. No-boundary thinking in bioinformatics research. *BioData Min*, 6:19–21.

Irwin JJ, Shoichet BK, 2016. Docking screens for novel ligands conferring new biology. *J Med Chem*, 59(9):4103–4120.

Kessler LG, Barnhart HX, Buckler AJ, et al., 2015. The emerging science of quantitative imaging biomarkers terminology and definitions for scientific studies and regulatory submissions. *Stat Meth Med Res*, 24(1). https://doi.org/10.1177/0962280214537333

Khoury MJ, Ioannidis JPA, 2014. Big data meets public health. *Science*, 346 (6213):1054.

Krane DE, Raymer ML, 2003. *Fundamental Concepts of Bioinformatics*. Harlow: Pearson.

Labov JB, Reid AH, Yamamoto KR, 2009. Integrated biology and undergraduate science education: a new biology education for the twenty-first century? *CBE Life Sci Educ*, 9(1):10–16.

Lopatto D, Hauser, C, Jones CJ, et al., 2014. A central support system can facilitate implementation and sustainability of a classroom-based undergraduate research experience (CURE) in genomics. *CBE Life Sci Educ*, 13(4):711–723.

Maloney M, Parker J, LeBlanc M, et al., 2010. Bioinformatics and the undergraduate curriculum. *CBE Life Sci Educ*, 9:172–174.

Merelli E, Armano G, Cannata N, et al., 2006. Agents in bioinformatics, computational and systems biology. *Brief Bioinformat*, 8(1):45–59.

Miskowski JA, Howard DR, Abler ML, Grunwald SK, 2007. Design and implementation of an interdepartmental bioinformatics program across life science curricula. *Biochem Mol Biol Educ*, 35:9–15.

Nasser T, Tariq RS, 2015. Big data challenges. *J Comput Eng Inf Technol*, 4:3.

Needleman SB, Wunsch CD, 1970. A general method applicable to the search for similarities in the amino acid sequence of two proteins. *J Mol Biol*, 48 (3):443–453.

Palsson B, 2016. *Systems Biology*. Cambridge: Cambridge University Press.

Russell SH, Hancock MP, McCullough J, 2007. Benefits of undergraduate research experiences. *Science*, 316:548–549.

Shaffer CD, Alvarez C, Bailey C, et al., 2010. The Genomics Education Partnership: successful integration of research into laboratory classes at a diverse group of undergraduate institutions. *CBE Life Sci Educ*, 9:55–69.

Tymann P, Congdon CB, Dougherty J, et al., 2005. Computer science and bioinformatics. In Steen LA (ed.), *Math & Bio: Linking Undergraduate Disciplines.* Washington, DC: Mathematical Association of America, pp. 75–82.

Venter JC, Adams MD, Myers EW, et al., 2001. The sequence of the human genome. *Science*, 291(5507):1304–1351.

Williams SM, Moore JH, 2013. Big data analysis on autopilot? *BioData Min*, 6:22.

Zhernakova A, Kurilshikov A, Bonder MJ, et al., 2016. Population-based metagenomics analysis reveals markers for gut microbiome composition and diversity. *Science*, 352(6285):565–569.

4 No-Boundary Course Developments

Joan Peckham, Bryan Dewsbury, Bindu Nanduri,
Andy D. Perkins, Donald C. Wunsch II,
and Yu Zhang

4.1 INTRODUCTION

Educators in K–12 have transformed standards, curricula, and pedagogical approaches to meet the demand of training students who will
become the next generation of problem definers and solvers. In the
United States, the Next Generation Science Standards (www
.nextgenscience.org) and the Core Curriculum State Standards (www
.corestandards.org) have both proposed a reduction in the number of
topics that are to be covered, but recommend more attention to
critical problem solving in settings where students are called upon
to apply the concepts and strategies that they have learned in many
other domains. With the focus on an integrated approach to teaching
and learning, the result is more interdisciplinary – if done well it will
provide young students with rudimentary no-boundary skills.

At post-secondary institutions, undergraduate and graduate programs now feature interdisciplinary courses and tracks that enable
students to prepare for emerging new areas of scholarship and employment. For example, bioinformatics and data science are two new
disciplines that have recently emerged from the fusion of multiple
other traditional disciplines, and are enhanced by new data-oriented
techniques. This is driven and supported by the demands of society,
industry, and government that provide incentives through external
funding for these educational programs and research projects.

Problems that do not fall neatly within disciplinary silos compel
students, scholars, and members of the workforce to work in no-
boundary groups, and this in turn demands better no-boundary
training at all levels. No-boundary skills represent an important

45

element of the *soft* or *essential* skills that industry has always desired in their employees. Due to the nature of the training that scholars usually receive, most are very focused on a particular domain. This sometimes makes it difficult to apply specialized skills to newly arising domains. However, to some extent the very best scholars have always applied deeply developed problem-solving approaches to new problems, and frequently do this with a team of differentially trained scholars. Problems such as the development of the nuclear bomb, the sequencing of the human genome, and the analysis of dynamic earth-bound systems to determine the impact of human behaviors on the ecology of the planet have all required no-boundary teams and approaches.

The purpose of this chapter is to outline a few first efforts to provide learning environments in which students learn no-boundary problem-solving skills. One additional outcome is that the meta-cognitive activities and reflections of the students while engaging in these learning environments also provide meaningful input to the formulation of the no-boundary approach. In this chapter we outline the experimental no-boundary classes that we have developed and reflect upon the challenges and outcomes.

4.2   RELATED RESEARCH

No-Boundary Thinking (NBT) works in areas where traditional inter-disciplinary approaches do not work well. These areas contain long-standing biological and chemical challenges such as the lung cancer prognostic signature problem or the problem of protein structure prediction. The traditional interdisciplinary approaches study these problems in two ways: (1) focusing the scientists on data gathering and directing the computational specialists to perform the data analysis; and (2) interpreting the discovered correlations of large datasets based on hypotheses. These hypotheses are based on a combination of preliminary and sometimes informal data analyses that are well grounded in the research team's intellectual knowledge and scholarship.

However, these long-standing problems have gone through long-term study with no significant breakthrough or advancement.

For example, the problem of identifying lung cancer biomarkers through analyzing large collections of genomic data has been studied for more than 10 years; there is still no gene signature ready for clinical use. This makes us rethink the fundamental problem definition in order to help patients on the clinical side. Instead of depending only on data and software tools as a connection between scientists and computational researchers, NBT provides a connection by enabling collaborative problem definition or redefinition of the scientific problems to address the scientific challenges. This brings a new approach to interpretation in data-driven research by encouraging and fostering the application of many disciplinary world views at the stage of defining the problem or hypothesis to be tested by the experimental method (No-Boundary Thinking Conference 2015). Therefore, the no-boundary approach is not just "deep" interdisciplinary research or a rigid addition of software and domain data. It is an infrastructure for computational science to reach beyond each discipline's boundaries, to attract the needed variety of researchers and resources, and to ultimately develop effective solutions. Table 4.1 highlights the features of NBT by comparing it with traditional interdisciplinary research.

It is with this framework and NBT goals in mind that we all created our first no-boundary classes intended to educate the next generation of scholars to address the most difficult problems of our and their times.

4.3 WHAT'S THE BIG IDEA? AN NBT APPROACH TO SOLVING COMPLEX SOCIAL CHALLENGES; SPRING 2016–2018, UNIVERSITY OF RHODE ISLAND, BRYAN DEWSBURY

4.3.1 The Nature of the Class

"What's the big idea" (WTBI) is an undergraduate course that has been framed around the NBT conceptual framework. Every time the course

Table 4.1 *The differences between No-Boundary Thinking and traditional interdisciplinary research*

Criteria	Traditional interdisciplinary research	No-Boundary Thinking
Connection between scientist and computational researcher	Connected by data and software tools	Connected by collaboratively defining problems
Data interpretation	Validating hypothesis	Multidisciplinary world views in problem and hypothesis definition
Overall	Congregating experts to solve problems	Integrated infrastructure for research problems, disciplinary expertise, and resources

is offered, a pertinent complex social challenge is chosen, the solution to which requires thinking broadly across several disciplines. The students spend the entire semester in groups developing the skills necessary to address the challenge, culminating in an oral presentation and written manuscript suggesting potential ways forward. The approach used in the course forces students to think and reason on content beyond their major, but to also apply their discipline-specific critical thinking skills to a broader, more complicated situation. In so doing, students should ultimately recognize that the problems of this world do not necessarily fall neatly along disciplinary lines, and thus future problem solvers need to have the ability to draw from multiple areas of scholarship to achieve solutions. In this vein, the WTBI course is considered to be a first-principles approach to problem solving. Regardless of discipline, the hope is that students understand that intellectual progress comes from holding oneself accountable to evidence-based reasoning, and a commitment to getting the facts right. To achieve this, students need to develop skills that would allow them to carefully parse the dimensions of a problem, and work as a team toward a pragmatic solution.

The course has a maximum enrollment of 16 students, who are divided into groups of four on the first day of class. The groups are carefully constructed to ensure that the students are from different majors. The first few weeks of the course are dedicated to teaching skills associated with sourcing peer-reviewed evidence, oral communication, and identifying the nature and scope of a social challenge. Understanding the challenge typically involves identifying the different areas of scholarship students need to engage with before problem solving can begin. Once these areas are identified, each group conducts a literature review on the area and leads a seminar-style class discussion. After the area-specific discussions, each group prepares a manuscript and a final presentation that encompasses the information received from the other groups informing a pragmatic way forward on the issue. In this final presentation each group highlights the NBT aspect of their approach.

4.3.2   Challenges and Benefits

The main challenge with a course like this is disabusing students of the notion that their intellectual potential is limited to the artificial boundaries set up by their major. Students typically have a natural fear of delving into the details of disciplines that seem academically distant to their chosen area of study. Proper scaffolding of the goals of the class can help alleviate this fear.

The benefits of the class are profound. Students who are primed to frame and discuss particular issues across disciplines are able to come to terms with the complexities of social challenges, and frame small-scale solutions that might address the original problem. After launching a new general education program, the University of Rhode Island (URI) has generously supported the development of new "Grand Challenge" courses. These courses are meant to be interdisciplinary in nature, and at the very least expose students to some of the most pressing complex issues of our time. "What's the big idea" fit seamlessly into the Grand Challenge program. Beyond merely exposing students to the issue, WTBI provides students with a framework to think about and address the issue.

4.3.3   Learning Outcomes

The overall goal of this course is for students to be able to understand the complexity of today's social challenges, and to adopt an effective and efficient mechanism for addressing those challenges. To this end, students in this course are expected to

- frame a complex issue using an NBT approach;
- contextualize their disciplinary expertise within potential solutions;
- apply progressive pedagogical approaches to explaining disciplinary fundamentals;
- apply the principles of teamwork to promote effective group work;
- communicate disciplinary understandings as well as solutions to complex issues;
- apply performance critiques to improve information communication; and
- provide critique to information delivery.

In spring 2016 WTBI focused broadly on the topic of climate change. The identified areas of scholarship were politics, history, biology, economics, and sociology. After detailed presentations on each area the groups crafted a small-scale solution to a localized climate change problem they identified. The presented solutions showed attention to detail, carefully thought-out analyses especially with respect to the multitude of factors, and a great appreciation for the ethical considerations that need to be given when solutions of those magnitudes are proposed.

4.3.4   Reflections

Teaching this course was a growing experience for the instructor's pedagogy, since the factor that drove its seeming success was the strength of the dialog within the classroom community. Student feedback suggested that though initially fearful, students appreciated the growth in confidence they experienced in discussing topics outside of their normal subject area. We have developed a pre- and post-assessment reflection assignment to qualitatively capture the degree to which this paradigm has occurred. The results of this analysis are forthcoming. We have also developed a similar version of the course to be offered as a study-abroad experience at the University of Cape Coast (UCC), Cape Coast, Ghana with funding support from the Office of the Provost at URI. For that version of the course, the identification of the challenge will be done in consultation with faculty from UCC, will enroll students from that institution, and will be about specific issues affecting the Ghanaian community. Our social challenge for the URI spring 2017 iteration was "America's opioid epidemic."

4.4 NBT INTEGRATED INTO MULTIPLE EXISTING COURSES; FALL 2015 TO SPRING 2017, DONALD C. WUNSCH II

4.4.1 *The Nature of the Classes*

Missouri University of Science and Technology (Missouri S&T) is the primary technological university for the state of Missouri. Starting salaries are the highest of any university in the state, and the placement rate is 93 percent. The institution has one of the nation's largest career fairs, with 307 employers participating last fall. Required experiential learning and real-world engagement are essential aspects of Missouri S&T's unique education environment. Undergraduate research is a core component that provides students with out-of-classroom learning opportunities to apply knowledge to real-world problems. Students participate in 18 student design teams, including national champion solar car and human-powered vehicle teams. Missouri S&T maintains a very engaged Engineers Without Borders organization on campus, with four groups and a high percentage of women participants and leaders. This rich array of hands-on learning opportunities contributes to the development of successful students who are ready to have an immediate positive impact when entering the workforce upon graduation.

Many graduate students are enrolled through Missouri S&T's extensive online and distance offerings; 761 of the 1929 graduate students participate in distance education. The Global Learning division provides a variety of credit and noncredit courses, seminars, conferences, and summer programs. In 2016 Missouri S&T was ranked in the top 10 nationally for "Best Online Graduate Engineering Programs" by *US News & World Report.*

Missouri S&T's College of Engineering and Computing offers 17 undergraduate degree programs and 10 minors across nine departments. All departments offer Ph.D. degrees, many with multiple specialties. Outside of engineering, many other disciplines are also offered through the doctorate, and even "service departments" have sufficient resources and interest to offer quality undergraduate degrees.

The courses receiving NBT integration at Missouri S&T are: Digital Logic Design (the introductory course in Computer Engineering) and graduate courses in Computational Intelligence, Markov Decision Processes, and Adaptive Dynamic Programming. Plans are in place to create a new course, Advanced Computational Intelligence, which will also receive NBT integration.

4.4.2   Challenges and Benefits

The aforementioned courses are all well-suited to NBT; each is multidisciplinary and every semester students enroll from multiple departments.

Challenges: The diversity of student backgrounds can be a two-edged sword. Some students are well prepared for one aspect of NBT but other aspects require significant learning of new concepts or rethinking of assumptions. Communication can also be a challenge. Furthermore, embracing NBT can require a greater degree of intellectual independence than standard class formats (book, lecture, homework, exams). For the graduate courses mentioned, the grades are primarily based on projects.

Benefits: Introducing NBT is a superb vehicle for developing the habits necessary for life-long learning. It is also conducive to collaboration and engagement with original literature. In Digital Logic Design, students learned to integrate computer engineering ideas with mathematical logic, graph theory, and optimization, and to appreciate the strengths of the several different majors who take the class. The graduate courses all received presentations on NBT as well as application problems in NBT, especially bioinformatics.

4.4.3 Learning Outcomes

Digital Logic Design students were able to optimize logic designs to fit any applications needing logical mappings or finite-state machines. At the small scale they know how to solve these by hand, and at the large scale they completely understand the algorithms they are applying.

The graduate students all learned to assimilate literature from outside their disciplinary boundaries. Each student also completed an NBT project applying new learning methods covered in the course (e.g. Xu and Wunsch 2009; Sieffertt and Wunsch 2010; Lam et al. 2015) to their own NBT problem. They also had to make multiple presentations in conference-style format.

4.4.4 Reflections

The project helped by explicitly identifying the virtues of NBT in subjects that naturally lend themselves to the concept. This impressed on students that knowledge, and pursuit of it, is actually one thing. The boundaries that we impose can be helpful to organize people into manageable units, but intellectually they are an illusion (Newman and Turner 1996). It is very helpful to have a community of people who recognize this reality.

4.5 AN NBT APPROACH TO HEALTH DISPARITIES: A ONE-CREDIT
COURSE FOR UNDERGRADUATE AND GRADUATE STUDENTS;
SPRING 2016, MISSISSIPPI STATE UNIVERSITY, BINDU
NANDURI AND ANDY D. PERKINS

4.5.1 The Nature of the Class

A trial special topics course was offered at Mississippi State
University, titled "A No-Boundary Approach to Health Disparities,"
during the spring 2016 semester, with the intention of using the
lessons learned to design a permanent course. This was a split-level
course, enrolling both graduate and undergraduate students. In the
spirit of other "Grand Challenges" courses, we sought to apply NBT
techniques to carefully define and address the problem of minimizing
health disparities, particularly within Mississippi. The National
Institutes of Health (NIH) definition of health disparity refers to
"differences in the incidence, prevalence, mortality and burden of
diseases and other adverse health conditions that exist among specific
population groups in the United States" and encompasses disparities
related to socioeconomic status (Carter-Pokras and Baquet 2002). In
2013, racial minorities comprised 42.4 percent of Mississippi, compared to the national average of 37.0 percent, which means that
without intervention, even if rates of sickness do not increase, the
state of Mississippi will only become sicker by virtue of a growing
minority population that is disproportionately impacted by unfavor-
able and preventable social determinants of health (El-Sadek et al.
2015). The challenge of addressing health disparities in the nation,
and in Mississippi in particular, requires data science and integration
of various aspects of science, including socioeconomics, and is an
ideal framework to introduce students to NBT. We designed the
one-credit hour seminar course such that each weekly meeting would
begin with a student-led discussion of a scholarly article or news
report addressing differences in health status or outcomes between
various groups of people, followed by an NBT-focused discussion.
Students wrote brief reflections on these discussions. The first two

class meetings led by the instructors introduced the concepts of big data, NBT, and health disparities. Students were allowed to choose the papers on which they wished to lead discussion, to allow the course content to be driven by student interests. We also kept the format of student-led discussion flexible. This flexibility allowed the students to choose a PowerPoint presentation, to share a paper of interest to read with the rest of the group a week before the class to allow a roundtable discussion during the class, or to role play group discussion where all students and instructors assumed the role of an important person who is invited to bring forth their expertise for applying NBT to mitigate a health problem. A final class meeting was reserved for a guest presentation by a campus expert on community engagement to promote health equity, to illustrate some of the practical aspects of trying to address health disparities in Mississippi at the grassroots level.

4.5.2 Challenges and Benefits

This course was not required as part of any degree program, and being a one-credit hour course would not wholly satisfy any degree requirements. Therefore, the main challenge of teaching the course is recruiting students to enroll. We contacted colleagues in various departments in the computational and life sciences, who discussed the course with their advisees. We found that primarily students who enrolled expected to find the course useful in their scientific training and future research. With Mississippi State University being home to the state's only College of Veterinary Medicine, and the course taught cooperatively by faculty from Computer Science and Engineering and Veterinary Medicine, students were primarily from the basic sciences or clinical veterinary programs.

The benefit to participation in the class for undergraduate or graduate students is the ability to lead a discussion based on their idea and looking at it from different perspectives, as opposed to presenting a research paper or project in a more traditional format for journal club or scientific presentations. Beyond the introductory sessions led by the instructors, the democratic approach to the discussion erases the implicit boundaries based on experience and allows everyone to participate as an equal and tries to bring consensus to the discussion where possible. The selection of the broad topic of health disparities, the flexibility in the selection of topics, and the format of student-led discussion facilitated the overarching goal of the class, which is to foster critical thinking, especially in a democratic group setting that is necessary for addressing grand challenges in science and society.

4.5.3 Learning Outcomes

The intended three primary learning outcomes for the course are:

1. Students will become familiar with the problem of health disparities within the United States and particularly in Mississippi.
2. Students will apply the NBT approach to problem definition and problem solving to the topic of health disparities in Mississippi.
3. Students will learn to work with those from outside of their discipline to define and solve large-scale scientific problems.

We realize that learning outcome 3 is largely dependent on the composition of the class, dictated by enrollment. Class participants were all from biology backgrounds. However, their specific focus within this broad discipline, and their individual training, ensured that there was enough diversity and disparity in the team that we had to bridge during class discussions.

Actual learning outcomes: In this first offering of the course, students selected a wide range of papers that were obviously closely related to their interests. Topics included populations disproportionately affected by the Zika virus outbreak (Samarasekera and Triunfol 2016), the accessibility of healthcare in refugee populations (El-Khatib et al. 2013; Sharara and Kanj 2014), and the impact on livestock of a refugee crisis, and even factors contributing to infections in shelter animals (Colby et al. 2011). Even though we had intended to focus on human health disparities, particularly within Mississippi, students led the discussion toward what they found interesting. This was often related to veterinary medicine, or topics appearing in the news such as Zika virus.

An undergraduate student led a role play discussion focused on the impact of smoking tobacco on cardiovascular disease. Among the characters/roles assigned were a medical student, a psychologist, a sociologist, a politician, and a CEO of a tobacco company, with the student as the moderator. All participants had to investigate an assigned area of a topic and assume that role during discussion and share their insights to the problem. All participants unanimously agreed that this is probably the best method to demonstrate a practical application of NBT. This activity demonstrated the challenges in accomplishing a task when applying NBT.

Thinking outside one's discipline is not easy even when it comes to role play; each of us researched a problem from our own perspective and comfort zone with respect to scientific knowledge. The class composition clearly showed that there are barriers to communication even within a single discipline as we began to have conversations about practical solutions to grand challenges in science. In most discussions students had to be led to consider the NBT aspects related to the problem.

We also found that students wished they had enrolled in a course such as this earlier in their college career, before they had begun research, as they found NBT helpful in defining their research problems. Students also valued the democratic discussion that allowed them to meet professors and students outside of their discipline as equals, an opportunity most of them had not had previously. Although this was not intended, the small class size (five) was appreciated by the students, as they felt that this was conducive to a more 'personal', interactive engagement in the class. Based on the course description and reading the NBT paper (Huang et al. 2013), students indicated that they thought the course would be skewed toward a traditional bioinformatics class and appreciated that it ended up being more about thinking, communicating ideas, and learning about genomics and human health.

The presentation by an MSU expert in addressing Mississippi's disparity in nutrition at the end of the course was well received. Students appreciated learning about real-life examples of addressing disparities in Mississippi as a practical demonstration of people from different walks of life coming together to solve a problem.

4.5.4 Reflections

The first offering of "No-Boundary Approach to Health Disparities" was a success. Although the focus shifted from the intended health disparities in Mississippi to topics of student interest, all class discussions were informative, engaging, and ultimately encompassed elements of NBT thinking – maybe even due to this altered focus. We will continue to offer this course and are finding ways to advertise it to graduate coordinators, student advisors, and different interest groups on campus. Health disparities will remain the focus in the future, as this is a problem that requires an NBT approach. Student evaluations support our enthusiasm to build upon this initial offering and enhance the learning experience for students.

4.6 NBT COMPUTER SCIENCE COLLOQUIUM AT TRINITY UNIVERSITY: A REQUIRED COURSE FOR ALL UNDERGRADUATE CS MAJORS AND MINORS; FALL 2015 AND SPRING 2016, YU ZHANG

4.6.1 The Nature of the Class

Trinity University is widely recognized as the top private undergraduate university in the Southwest. Undergraduate research is central to the overall goals of Trinity University. Trinity's mission to provide students with a thorough understanding of principles and processes is accomplished not only through classroom and laboratory study, but also through engagement of students in active research. The Computer Science Department at Trinity University recognizes and endorses the university's overall goals. The design of the Computer Science Colloquium reflects this by aiming to:

- expose students to a variety of topics relevant to the field, including current research, professional opportunities, and ethics;
- provide a venue for research presentations by both faculty and students;
- provide a venue for outside speakers to present their research and industry experiences to students; and
- foster a sense of community among students and faculty.

4.6.2 Challenges and Benefits

The Computer Science Colloquium is a weekly 75-minute class required of all computer science majors and minors. It is offered every semester, but students only need to take four semesters of it and they can choose any semester to register for the course. The average class size is about 60 students each semester. The course has existed for five years, since before NBT was introduced. In the academic year 2015–2016 we set NBT in computing as a year-long special theme for the colloquium. We invited 16 guest speakers to present their research to the students. They came from Trinity University, the University of Texas at San Antonio, the University of Texas Health Science Center at San Antonio, and the Southwest Research Institute, with expertise in:

- biophysics
- bio-organic chemistry and biochemical engineering
- cognition and emotional disorders
- computational fluid dynamics in mathematical biology
- computer science
- cybersecurity
- epidemiology and biostatistics
- genetics
- mechanical engineering
- neuroscience
- nonparametric statistics
- robotics
- topology, geometric theory, and diffeomorphisms.

In spite of the different backgrounds of the speakers, they all use computers to do modeling and simulation in their research, or they collaborate with computer scientists to do it. During their talks they presented some common challenges and benefits of doing collaborative research with or in computer science.

Challenges: For real-world problems, especially those in the natural sciences, data collection is the heart of the research (Lynch 2008). But it is easy to be misled by the overwhelming amount of data available today. Moreover, quantity and quality are not the same, and finding reliable, relevant data requires well-balanced background knowledge on the part of researchers. When defining problems, no matter how big the data is, we need to understand where the underlying science challenge comes from; thus, experts from different areas need to work together to conceptualize, capture, and define the problems without disciplinary boundaries.

Benefits: The NBT colloquium encourages more young and talented students to pursue research and to do research in a well-established interdisciplinary environment. Multiple speakers expressed their desire to involve Trinity students as part of their research groups, exposing them to the intellectual excitement of various computational research problems, and thereby encouraging them to think creatively and independently and to develop the skills necessary to work on interdisciplinary projects.

4.6.3 Learning Outcomes

By the end of the year the class had held two contests: an NBT essay contest and an NBT logo contest. By participating in either contest students expressed their understanding of and interest in NBT in science (the essay contest) or in art (the logo contest). All participants presented their outcomes to the whole class (we also invited department faculty to attend the presentations). The faculty and the students voted and selected the four best essays and one best logo. The evaluation criteria for the essay contest are:

- Is the problem multidisciplinary (more than one discipline)?
- Does it fit into NBT?
- Are the data and the way they are collected reasonable?

Here are the results:

NBT essay winners

- The physical web: how beacon technology can add context to our world (Kylie Moden).
- Gene name disambiguation: a novel approach (Evan Cofer).
- Group theory: a computational approach (Andy Leeds).
- Using computer science to make campaigning more efficient (Cameron Hayes).

NBT logo winner

- The winner was Kat Fisher; her design is shown in Figure 4.1.

4.6.4 Reflections

The Trinity NBT Colloquium was a success. Students liked the broadness of the topics as well as the dynamics between computer science and other science disciplines. Through the class they realized that computer science and technology are often central to much scientific research. Computer science research requires not just domain experts, such as medical and environmental researchers, but also people with deep knowledge of math and logic who are able to develop algorithms and complex applications to facilitate scientific research.

FIGURE 4.1 Trinity NBT logo contest winning design.

4.7 NO-BOUNDARY RESEARCH SEMINAR: A ONE-CREDIT GRADUATE CLASS; FALL 2015–2017, UNIVERSITY OF RHODE ISLAND, JOAN PECKHAM

4.7.1 The Nature of the Class

The University of Rhode Island has created a graded graduate one-credit course entitled "No-Boundary Research Seminar." The class is taught each fall semester and attracts students from multiple disciplines, mostly from the sciences, and social sciences. To date it has been taught three times: fall 2015, fall 2016, and fall 2017. In this class students explore the nature of NBT and how it might transform the conduct of scientific scholarship and the scientific method.

The students read, discuss, and reflect upon articles about interdisciplinary and no-boundary engagement. Visitors who are engaged in no-boundary projects attend the class and provide their perspectives. Some present to the students and allow for questions, and others prefer to engage the students in activities to encourage them to formulate their own ideas about the nature of NBT. Some are URI professors and others are visitors from other institutions who also give a research seminar during class time, which attracts a broader audience. All are chosen as exemplars of no-boundary engagement. For example, Druschke is a scholar of writing, rhetoric, and environmental science (Druschke and McGreavy 2016), and a graduate of an NSF-funded interdisciplinary Ph.D. program (IGERT) in the Midwest.

While students arrive to class on the first day assuming that the instructor and visitors will define NBT for them, this is meant to be an active learning environment, so the students are provided a preliminary definition, exposed to a variety of approaches and opinions, and asked to join the no-boundary community to explore and define it.

Activities in the class range from short exercises in no-boundary problem definition in groups of individuals with multiple perspectives, to discussions of readings around inter-, multi-, and transdisciplinary and no-boundary scholarship and education. Individually, students are asked to write regular reflections on their engagement in the class and their newly formulated ideas about NBT. The final assignment is a group project in which the students define a no-boundary problem, and then report on the problem definition process through a presentation, and offer their final individual reflections.

4.7.2 Challenges and Benefits

The University of Rhode Island is currently moving toward no-boundary engagement in both research and education. A new undergraduate general education program that is intentionally integrative began in fall 2016. Instead of taking courses from a menu of traditional disciplines (mathematics, science, fine arts, social science, languages, etc.), students must cover 13 integrative areas such as: understand and apply STEM theories and methods, communicate effectively, develop and exercise global responsibilities, and explore multiple perspectives of areas of contemporary significance and their ethical implications (see web.uri.edu/advising/general-education-2016). Similarly, scholarship is taking a no-boundary turn with tenure-track cluster hires focusing on problems that span traditional scholarly silos – for example, big data and data science, neuroscience, and Islamic studies. Scholars in departments that have traditionally provided consulting services to other disciplines are now fully engaged as collaborators in research. To support this, the graduate school permits co-major advisors from different disciplines, and existing graduate programs are considering new interdisciplinary tracks.

All of these developments provide a supportive environment for this class. Challenges include recruiting students from disciplines where existing program requirements are tight. However, offering the class for one credit has helped with this. Another challenge mentioned above is moving the students from passive listening to active participation in exploring the different aspects of no-boundary engagement, but so far this has not been a serious challenge once the students are convinced they have "permission" and are expected to actively define and explore.

4.7.3 Learning Outcomes

Intended learning outcomes of the class include expectations that students will show they are able to:

- use no-boundary skills where needed in their own research and future careers;
- productively engage in the ongoing definition of no-boundary research and education;
- discover and articulate the important features of and processes enabling genuine NBT; and
- produce a written reflection that goes beyond reporting what they have observed and heard in class with a rhetorical piece in which they present an opinion or new idea and then support it.

4.7.4 Reflections

Students in the classes have embraced the activities and, through their reflections and written evaluation of the class, have contributed significantly to the broader efforts to understand the nature of NBT and what it takes to carry it out effectively. This strengthens the students' no-boundary skills, but also improves the class for the next offering. For example, effective no-boundary communication was a strong focus of the students in both classes. One year, one student suggested deeper exploration of the epistemology of interdisciplinary or no-boundary inquiry. Another introduced the class to visible thinking principles for deepening student understanding and configured an activity in which groups in the class could use visible thinking to enable more robust no-boundary problem definition. In future classes we will include these topics and materials for exploration (e.g. O'Rourke 2013; Project Zero 2017)

ACKNOWLEDGMENTS

Donald C. Wunsch: Partial support for this research was received from the Missouri University of Science and Technology Intelligent Systems Center, the Mary K. Finley Missouri Endowment, the

Lifelong Learning Machines program from DARPA/Microsystems Technology Office, and the Army Research Laboratory (ARL); and it was accomplished under Cooperative Agreement Number W911NF-18-2-0260. The views and conclusions contained in this document are those of the authors and should not be interpreted as representing the official policies, either expressed or implied, of the Army Research Laboratory or the US Government. The US Government is authorized to reproduce and distribute reprints for Government purposes notwithstanding any copyright notation herein.

Joan Peckham and Bryan Dewsbury would like to acknowledge that partial support for the development of their no-boundary classes at URI was provided by a subcontract to EAGER NSF Award, NSF – 1452211, Arkansas State University, PI Xuizhen Huang. Any opinions, findings, and conclusions or recommendations expressed in this material are those of the authors and do not necessarily reflect those of the NSF.

REFERENCES

Carter-Pokras O, Baquet C, 2002. What is a "health disparity"? *Public Health Rep*, 117(5):426–434.

Colby KN, Levy JK, Dunn KF, Michaud RI, 2011. Diagnostic, treatment, and prevention protocols for canine heartworm infection in animal sheltering agencies. *Vet Parasitol*, 176(4):342–349.

Druschke CG, McGreavy B, 2016. Why rhetoric matters for ecology. *Front Ecol Environ*, 14(1):46–52.

El-Khatib Z, Scales D, Vearey J, Forsberg BC, 2013. Syrian refugees, between rocky crisis in Syria and hard inaccessibility to healthcare services in Lebanon and Jordan. *Conflict Health*, 7(18). https://conflictandhealth.biomedcentral.com/articles/10.1186/1752-1505-7-18.

El-Sadek L, Lei Zhang L, Vargas R, Funchess T, Green C, 2015. *Mississippi Health Disparities and Inequalities Report*. Jackson, MS: The Office of Health Disparity Elimination and the Office of Health Data & Research, Mississippi State Department of Health.

Huang X, Bruce B, Buchan A, et al., 2013. No-boundary thinking in bioinformatics research. *BioData Min*, 6:19.

Lam D, Wei M, Wunsch D, 2015. Clustering data of mixed categorical and numerical type with unsupervised feature learning. *IEEE Access*, 3:1605–1613.

Lynch C, 2008. Big data: how do your data grow? *Nature*, 455:28–29.

Newman JH, Turner FM, 1996. *The Idea of a University*. New Haven, CT: Yale University Press.

No-Boundary Thinking Conference, 2015. Little Rock, AR, April 12–14. http://plantimg.uark.edu/nbt2015.

O'Rourke M, 2013. *Interdisciplinary epistemology: syllabus*. Michigan State University. https://msu.edu/~orourk51/860-Phil/Handouts/860F13-hdout.htm

Project Zero, 2017. Visible thinking. www.visiblethinkingpz.org/VisibleThinking_html_files/VisibleThinking1.html.

Samarasekera U, Triunfol M, 2016. Concern over Zika virus grips the world. *The Lancet*, 387. https://doi.org/10.1016/S0140-6736(16)00257-9.

Sharara SL, Kanj SS, 2014. War and infectious disease: challenges of the Syrian civil war. *PLoS Pathogens*, 10(11):e1004438.

Sieffertt J, Wunsch DC, 2010. *Unified Computational Intelligence for Complex Systems: Studies in Neural, Economic and Social Dynamics*. Berlin: Springer.

Xu R, Wunsch DC, 2009. *Clustering*. New York: IEEE Press/Wiley.

5 No-Boundary Thinking for Transcriptomics and Proteomics Big Data

Mariola J. Ferraro, Andy D. Perkins, Mahalingam Ramkumar, and Bindu Nanduri

5.1 INTRODUCTION

Ideal healthcare should provide prevention and treatment strategies in the context of individual variability. Achieving this goal through precision medicine that embodies personalized, predictive, preventive, and participatory patient care (that is, the practice of individualized P4 medicine) (Hood and Friend 2011) in the near future relies on the integration of genomics and big data using data science techniques. Completion of the Human Genome Project (Lander et al. 2001; Kim et al. 2014), in conjunction with continued improvements in genome sequencing technology, propel generation of big data that needs to be integrated with clinical data for precision medicine. High-throughput techniques, collectively termed as omics approaches, allow the measurement of gene expression (transcriptomics), protein expression (proteomics), and metabolic status (metabolomics). Integrated analysis of omics data in a systems biology framework, using mathematical modeling and applying data science techniques, is required for a comprehensive understanding of disease mechanisms and to provide insights into the physiological state of an organism or a cell. Precision medicine will be at the forefront of the public health sphere for the next few decades until the promise of genomic medicine that hopes to bring "base pairs to bedside" (Green et al. 2011) is fulfilled.

The road map for individual genomics for precision medicine will require a paradigm shift in how academic research is conducted, spanning education, training, data integration, ethics, regulatory compliance, and data security, to name a few. At the core of this conversation is the data. The collection, storage, and analysis of data to address a specific health problem, whether it is cardiovascular disease, diabetes, cancer, Alzheimer's, or the plethora of infectious and zoonotic diseases, will be as varied as the disease under investigation and will require building heterogeneous workflows to solve the problem. Understanding the complex disease etiology and development of treatment strategies for translating to the bedside will not be possible with the current mindset where researchers with expertise in specific disciplines work in silos and communicate only at the stage of data integration, which is often the last step of a study.

Data collection is often carried out in isolation, without any thought given to the ultimate overall goal. The No-Boundary Thinking (NBT) approach (Huang et al. 2013) that advocates a scientific dialog among individuals with varying expertise in a "discipline-free" manner at the problem-definition stage, is a pragmatic approach to leverage big data for precision medicine. Genomics big data as it pertains to understanding the molecular function of genes and proteins is the focus of this chapter. The purpose of this chapter is not to provide a comprehensive review of experimental methods and data analysis pipelines for transcriptomics and proteomics. We will briefly introduce these two functional genomics approaches to study gene and protein expression and discuss the application of the NBT approach to accelerate omics-based discovery for personalized medicine.

5.2   PROTEOMICS

Among the multitude of omics approaches that offer a snapshot of biological systems from different perspectives, proteomics provides a unique view of the cell that is applicable to various biomedical fields. Proteins, which are ultimately the products of transcribed genes, interact with one another to carry out a specific physiological response, thus regulating virtually all cellular processes. As such, proteins constitute attractive drug targets and often can serve as diagnostic markers or fingerprints of the physiological state of a cell.

Understanding the protein complement of the genome, also known as a proteome, is critical for understanding cell physiology.

Studies of protein function and regulation traditionally involve purification of individual proteins and biochemical studies of their structure and potential activity. Methods such as enzyme linked immunosorbent assays (ELISA), immunohistochemistry, and immunoblotting, which utilize antibodies for specific proteins, can provide information about the level of specific proteins in a cell or tissue. All these methods rely on the specificity of antibodies, which recognize certain epitopes of individual proteins. This protein–antibody interaction is then visualized as a signal by using various approaches. While these antibody-based methods of detection are useful, they constitute reductionist approaches that are not at the genome scale. Mass spectrometry-based analysis of the proteome constitutes the study of proteins on a global genome scale, generates big data, and is a major focus of this chapter. This approach can provide information about the expression of hundreds and even thousands of proteins simultaneously, and can also be applied to map protein interactions, mutations, and post-translational modifications.

Several mass spectrometry-based approaches are described in the literature that help study distinct characteristics of proteins. These strategies include bottom-up and top-down approaches. In the bottom-up proteomics approach, often referred to as shotgun proteomics (Zhang et al. 2013), proteins are digested with an enzyme such as trypsin or chymotrypsin and the obtained peptides are analyzed by a mass spectrometer. If this approach is combined with a separation method, such as liquid chromatography, it enables analysis of multiple proteins at the same time.

The dynamic range of proteins can be seven orders of magnitude, including one copy per cell to ten million copies per cell (Zubarev 2013). Therefore, deconvolution of the peptides by chromatographic techniques enhances the proteome coverage. In shotgun proteomics, data-dependent analysis is used, where first the peptides are analyzed at a full spectrum of mass/charge range, which is then followed by separation of precursors and analysis of their fragmentation spectra. This fragmentation most often is achieved by utilizing collision-induced dissociation (CID) and a type of CID method called higher-energy collisional dissociation (HCD), although other fragmentation methods are also increasingly used, such as electron transfer dissociation (ETD). Data-dependent analysis is utilized to provide information about the amino acid sequence of each analyzed peptide. The masses of analyzed parent ions and individual fragment ions are compared to the *in silico* map of the digested proteome, which finally yields the peptide sequence and matches it to individual proteins. This information can be further complicated when a peptide is not unique to just one protein, but is shared among several orthologs or related proteins. In this case, other evidence is required to prove a presence or abundance of this protein in a sample. Algorithms that match experimental mass spectra with theoretical mass spectra are central to peptide and protein identification for this untargeted bottom-up approach. Significant change in the relative abundance of proteins in two different conditions is carried out post-identification using different imputation methods and statistical methods (Aebersold and Mann 2016).

Targeted bottom-up proteomics combines the use of trypsin digestion of proteins and selected reaction monitoring (SRM) by a triple quadrupole (QQQ) instrument (Picotti and Aebersold 2012) in which CID is used to increase selectivity. The SRM workflow involves monitoring a particular fragment ion of a selected precursor ion using two mass analyzers and recording the intensity value over time. Several SRM precursor–fragment ion pairs can be monitored in the same run in the multiple reaction monitoring (MRM) method. This approach can be successfully used to analyze a specific list of peptides with a known mass to charge ratio, and it can help fragment and quantify several peptide targets during the same run. If the peptides monitored by the MRM method are sufficiently accurate to distinguish a protein of interest, compared to shotgun proteomics that might not detect a target of interest, the MRM approach is preferable, as it monitors a target protein in a complex sample at all times based on specific mass to charge ratios of its peptides. MRM methods are considered to outperform traditional western blotting techniques, which are commonly used to analyze differences in protein levels between samples. Compared to western blotting techniques, which are limited by antibody availability and/or lack of specificity, MRM can provide absolute quantitative values that are superior and accurate (Aebersold et al. 2013).

Top-down proteomics provides the means to detect protein degradation products and post-translational modifications, and can be used to analyze stoichiometry of proteins in a mixture (Barrera et al. 2009; Vorontsov et al. 2016). In this technique, undigested, intact proteins are analyzed directly by a mass spectrometer, which is followed by fragmentation of intact proteins and analysis of the mass to charge values of the product ions. There are certain advantages to the top-down approach – for instance, it can be useful for detection of degradation products, which might not be possible by the typical shotgun proteomics approach. Also, this method is thought to be beneficial for detection of specific post-translational modifications (Toby et al. 2016).

Top-down proteomics requires the use of high-resolution mass spectrometry for the resolution of individual isotopic peaks of proteins. Resolving isotopic peaks is critical for determining the charge of ions, which is necessary for estimating protein mass. Isotope resolution in this approach is due to the fact that intact proteins are large and have multiple charged amino acid residues. However, sufficient progress in intact protein separation methods, fragmentation strategies designed for intact protein analysis, and data analysis approaches for top-down proteomics could facilitate application of this technique to obtain more information about the protein isoforms while avoiding the complexity of peptides.

While the accumulation of a significant amount of peptide or intact protein fragmentation data can still be problematic due to the cost and time associated with comprehensive proteome analysis, other challenges of classical shotgun proteomic analyses are designing appropriate experimental controls and data analysis. An extensive study in 2009 analyzed the same sample consisting of 20 different proteins in 27 different mass spectrometry labs, followed by secondary data analysis by an independent lab (Bell et al. 2009). Interestingly, this study concluded that although most of the laboratories generated a mass spectrometry data of very high quality, the bioinformatic approaches were a limiting factor in the correct identification of the proteins.

The description of the human proteome was reported in 2014 (Kim et al. 2014). In this study, 30 human tissues/cell types from three different individuals per sample type were analyzed by high-resolution Fourier-transform mass spectrometry. Over 17,000 proteins were identified, which constitutes approximately 84 percent of the known protein-coding genes. Among the identified proteins, almost 200 were novel and represented genes previously assumed to be noncoding regions in the genome. Another study (Wilhelm et al. 2014), which interestingly appeared in the same volume of *Nature*, used newly generated data as well as publicly available proteomics data, leading to the identification of 18,097 proteins of the annotated 19,629 human genes (based on SwissProt database content), which accounted for 92 percent of human genes. While these studies demonstrate the current status of the human proteome, which is valuable in itself, they are also critical for integrative studies with other types of omics data. An integrative personal omics profile, combining the use of genomic, transcriptomic, proteomic, metabolomic, and autoantibody profiles from a single individual, collected over a period of 14 months, identified several medical risks in the studied person, including type 2 diabetes. Moreover, it also showed the dynamic nature of the changes in the data types collected, and the dependence on the physiological state of this individual (Chen et al. 2012).

The sensitivity of current mass spectrometers has enabled the identification of a large number of proteins simultaneously and could allow this technique to completely replace most of the standard one-molecule-based methods for analyzing protein expression in an unbiased manner under different physiological conditions. Successful implementation of this approach would result in rapid progress in the biomedical field as it pertains to our understanding of post-transcriptional genome regulation and could ultimately contribute to precision medicine. However, there are still major caveats with this technology that render traditional biochemical techniques attractive and possibly only alternatives in some cases. For example, in comparison to gene analysis, proteins cannot be amplified and the less abundant proteins and polypeptides might be below the level of detection for most mass spectrometers. Also, the existence of multiple isomers and post-translational modification can limit the detection and confident quantification of proteins.

5.2.1 Proteomics: Challenges and Future Perspectives

Advances in technology are resulting in mass spectrometers that are more sensitive and faster. Alongside the gains in technology, there are constant improvements in data analysis, integration, and management. A significant advance in technology is the development of the Orbitrap mass spectrometer, which utilizes Fourier-transform for accurate mass measurement. These instruments also provide high resolving power and relatively fast sample analysis, and all at a low cost and comparatively low complexity (Perry et al. 2008). Other instruments frequently used for shotgun or discovery proteomics are quadrupole ion trap (Paul trap), time of flight (TOF) and FT-ICR mass spectrometers (Table 5.1). However, the Orbitrap is one of the most versatile mass spectrometers currently utilized by biomedical researchers.

Despite the progress made in technology that has seen massive improvements in mass spectrometers, and the increasing number of high-throughput studies of proteomes, there is no corresponding increase in the number of diagnostic biomarkers, drugs, or other products. In fact, from over 150,000 papers describing putative biomarkers discovered by such technologies as proteomics or microarrays, only a small fraction of fewer than 100 has been

Table 5.1 *Sensitivity, mass accuracy, and resolving power of different mass spectrometers (Bereman et al. 2008).*

Instrument	Sensitivity (g)	Mass accuracy (ppm)	Resolving power (FWHM)
Quadrupole ion trap	10–15	50	10,000
TOF	10–12	3	20,000
FT-ICR	10–12	1	1,000,000
Orbitrap	10–15	2	100,000

validated for their potential use in clinics (Poste 2011). Some of the biggest challenges in translating proteomics data into clinics are related to improper study design, selection of appropriate cohort, lack of controls, specific bias in patient choice, inappropriate conditions during sample collection or storage, application of incorrect statistical and data analysis methods, and selection of inappropriate validation methods (Wehling 2021). To address some of these issues, standards regarding expected phases of biomarker discovery and validation have been developed (Pepe et al. 2001; Plebani 2005). Stringent guidelines are being established as "best practice" models to ensure appropriate biobanking (Poste 2011). Finally, a collaborative environment between clinicians, engineers, basic science researchers, biostatisticians, and other scientists is highly recommended (Pavlou et al. 2013). The proteomics community should come together to propose guidelines similar to the ones proposed for proteomics-based biomarker discovery and additional proteomics-based applications in studying diseases. For example, in proteomics-based studies of infectious disease, which involve highly complex host–pathogen interactions, attention should be given to identifying the optimal biological model, development of an appropriate predictive model based on the data, and also the choice of validation approaches. Validation could include the use of mutant or knockout animals, which again requires close collaboration between different stakeholders (Aderem et al. 2011).

The diversity of data acquisition methods and data analysis workflows, the need for comprehensive metadata, and data standards discussed in this section do not even include another important application of proteomics, which is in the area of improving genome annotation by proteogenomics (Nesvizhskii 2014). In proteogenomics, peptide-level data combined with DNA sequence and additional RNA-Seq-based data can identify potentially novel small protein-coding genes missed in the initial annotation, identify splice junctions, provide evidence for long noncoding RNA, etc., which ultimately enhances the information content in the genome sequence. We have developed proteomics analysis pipelines, conducted proteogenomic mapping, and utilized expression proteomics and quantitative proteomics for our research. Having the necessary expertise to maximize the potential of a single omics method, such as mass spectrometry for protein analysis, with its myriad rapidly evolving biological applications, is not possible without a collaborative approach, even at an individual research lab scale. The prospect of integrating proteomics, which provides a glimpse of post-transcriptional, translational, and post-translational regulation of genome expression, with other omics data requires expertise and multidisciplinary collaborative teams on a much larger scale. In fact, this need for partnerships for harnessing the power of 'big biology' for solving healthcare challenges was cited as one of the lessons learned from the Human Genome Project (Green et al. 2015).

5.3 TRANSCRIPTOMICS

Transcriptome analysis is an ideal illustration of the potential of interdisciplinary research and NBT. The analyses described here are dependent upon a wide range of expertise in terms of defining both the biological and computational problem. In a standard RNA-Seq approach, sequence reads must be aligned to a reference genome or a transcriptome assembled *de novo*. This has been an important topic in computer science since the development of the Smith–Waterman algorithm (Smith and Waterman 1981). Common methods for differential expression (Anders and Huber 2010; Robinson et al. 2010;

Seyednasrollah et al. 2015) are dependent upon statistical approaches, and clustering methods draw expertise from both of these areas, as well as theoretical computer science and operations research. However, too many studies are conceived and carried out without involving computer scientists and statisticians in the problem definition and experimental design.

Examination of the entire transcriptome of an organism has been common since the earliest serial analysis of gene expression (SAGE) studies (Velculescu et al. 1995) and the development of high-density oligonucleotide arrays, or microarrays (Lockhart et al. 1996). These approaches provide a "snapshot" of transcription activity by allowing the quantification of the cellular concentration of messenger RNAs (mRNAs). As high-throughput sequencing technologies have become available and more advanced, RNA-Seq has emerged as the preferred method for gene expression analysis in many domains. RNA-Seq makes use of these high-throughput sequencing technologies to sequence complementary DNA (cDNA) that has been reverse-transcribed from mRNA transcripts. The use of high-throughput sequencing technologies ushered transcriptomics into the big data era, with many billions of sequence "reads" commonly generated by each gene expression study. As of early 2017, the Sequence Read Archive (SRA) (Leinonen et al. 2011), housed at the National Center for Biotechnology Information (NCBI), contained over 10 petabases of sequence data and even more petabytes of data when quality and other information stored is considered.

The attractiveness of the RNA-Seq approach is at least partly due to its applicability in the absence of the sequence of individual transcripts, and the value added by other analyses that can be performed in addition to the standard quantification of gene expression. Data generated with the RNA-Seq approach can be used to identify novel transcripts or isoforms (Roberts et al. 2011), refine genome and transcriptome structural annotation (Lockhart et al. 1996; Li et al. 2011), and uncover single nucleotide polymorphisms (Piskol et al. 2013; Lopez-Maestre et al. 2016) and other sequence variation within transcribed regions. Furthermore, transcriptomic sequence reads can be assembled using methods similar to those for genome assembly (Grabherr et al. 2011; Xie et al. 2014) to produce a representation of the entire transcriptome, and a scaffold onto which sequence reads can be aligned for quantification.

The most common use of gene expression measurements is in differential gene expression analysis. In this type of analysis, gene expression levels are compared across two or more groups – for example, a control and treatment group. The goal here is to identify transcripts that show a significant difference in abundance between the groups, with the expectation that transcripts identified are related to the treatment being studied. In addition, the concept of "guilt-by-association" (Wolfe et al. 2005), which suggests that genes with similar expression profiles tend to be functionally related, has led to the wide-scale use of clustering methods on gene expression data and the generation and analysis of gene co-expression networks. In these analyses, genes exhibiting similar expression patterns over a number of samples, as determined by some similarity or distance metric, are grouped together using a selection from the wide range of available clustering algorithms (Jay et al. 2012). Functional annotation for those genes with unknown or limited functional information can then be inferred from that of other cluster members. Likewise, gene co-expression networks are constructed by linking genes with similar expression profiles. These networks can then be mined for information about which genes may be participating in similar functions, or compared across groups of samples.

While the term "transcriptome" has traditionally been applied to the repertoire of mRNA transcripts within an organism, this definition has expanded to accommodate the increased understanding of microRNAs (miRNA) and other regulatory and noncoding RNAs (Morris and Mattick 2014). Similar methods as RNA-Seq can be used to investigate miRNAs, for example. As these data are generated, they can be integrated with gene expression data to provide a more complete picture of the transcriptome. One approach to this integration is to incorporate measurements of these RNAs into gene networks to elucidate the functional role of, and interactions between, various components of the transcriptome.

5.4 BIG DATA AND RESEARCH TEAMS

The genomics scientific community embraces data sharing, and this is now becoming a common practice in mass-spectrometry-based proteomics (Martens and Vizcaino 2017). A number of public repositories, including PRIDE Archive (Vizcaino et al. 2016), serve as centralized, standards-compliant repositories for proteomics data. These resources pave the way for integration of proteomics data with other types of functional genomics data. In the biological context this integration has to account for differences in the timescale of genome expression, such as rapid (activation of enzyme activity) versus long-term response (transcriptional). In the context of bioinformatics, integration requires data interoperability, differences in the measurement scales, varying signal–noise ratios inherent to each analytical technique employed, and reconciling multiple algorithms used within and across different omics platforms. Although data and metadata standards set forth by the scientific community address some of these issues while ensuring transparency, there is a need for active research in bioinformatics and computational biology to address this barrier to data integration. A successful example of integrated resources that provide insight into basic biology is the development of model organism databases (Tang et al. 2015). The premise for developing model organism resources is based on decades of research that clearly shows the conservation of basic operating principles of certain biological processes between humans and different organisms, such as animals, plants, or microbes, that are easy to access, maintain, and manipulate. Model organism databases such as the Saccharomyces Genome Database and the Mouse Genome Database are focused on providing well-curated organism-specific, comprehensive, integrated biological information. This concerted effort often results in data analysis methods and publicly available resources that can be used for rapid biological discovery. In addition to model organism databases for proteomics research, a number of reference proteomes are available in Uniprot, a protein-centric bioinformatics database (UniProt Consortium 2017). The current release of Uniprot (as of March 20, 2017) has 8139 reference proteomes that show broad taxonomic representation, including bacteria, viruses, archaea, and eukaryota. A reference proteome represents a selection of a complete proteome – that is, a set of proteins expressed in an organism with a genome sequence from a select organism that is manually and computationally curated. The reference proteome approach is a pragmatic solution to reduce the dimension of the available proteomes in a meaningful and biologically relevant manner.

Yet another collective, collaborative, often multi-institutional, team-based approach to solving biological problems, whether it is in the area of data generation, development of data standards, or synthesizing knowledge in a specific area of biology, that has proven to be successful is the prevailing trend of conducting consortium biology. "Consortium biology is a research program led by a complementary set of laboratories or institutions, all working towards a common and well defined goal," as defined by the Immunological Genome Project (Benoist et al. 2012). The magnitude and/or requirement for multidisciplinary skills mandate that consortium biology that is not just a collation of independent projects that focus on different aspects of the same research theme. The Human Genome Project is an excellent example of consortium biology (Lander et al. 2001; Kim et al. 2014). The ENCODE Consortium (2012) is the encyclopedia of functional DNA elements in the human genome. Similarly, efforts to catalog common patterns of DNA sequence variation among individuals to help determine individual susceptibility to disease were the focus of international consortium projects such as HapMap (International HapMap 2003; Thorisson et al. 2005) and the 1000 Genomes Project (Auton et al. 2015) that generated an invaluable comprehensive view of global human variation. The ProteomeXchange Consortium for coordinated submission of mass spectrometry data (Deutsch et al. 2017), the

UniProt Consortium (UniProt Consortium 2017) (which curates protein sequences and annotates their function), the Consortium for Top-Down Proteomics (for accelerating analysis of intact proteins by refining and validating top-down proteomics data) (Dang et al. 2014), and the Clinical Proteomic Tumor Analysis Consortium (Ellis et al. 2013) (which applies proteomics to annotate tumors from The Cancer Genome Atlas program) are all consortium-scale projects in the field of proteomics that focus on different aspects of proteomics, spanning data collection through application to different tumors.

5.5 PROTEOMICS AND TRANSCRIPTOMICS BIG DATA MANDATES NBT

Big data is pervasive in all disciplines of science and society. The definition of big data keeps evolving, as does the data itself. The most common description of big data is exemplified by the Vs: volume, velocity, and variety (Laney 2001). Volume refers to the sheer size of data that needs management, velocity refers to the pace of data generation, and variety captures the heterogeneity of data formats, structures, and semantics. Over time additional Vs – such as veracity, visualization, and value (Sowe and Zettsu 2014) – were needed and were included in the definition. In the context of omics data for precision medicine, the promise of big data relies on the ability to harness information in novel ways to generate insights for individualized treatment. Therefore, leveraging big data for gains in human health will ultimately depend on how we process and analyze the data, moving beyond the data generation phase. Consortium science in biology has generated big data. While the contribution of these projects is critical for the current advances in the area of precision medicine, lessons learned from these projects mandate new strategies for addressing the challenges of bridging the genotype–phenotype and bench–bedside gaps that will convert investments in science to societal benefits.

Completion of The Cancer Genome Atlas project that genetically profiled 10,000 tumors and generated 20 petabytes (10^{15} bytes) of data showed that the tumor landscape is highly complex, with very few mutations identified as being drivers. Although it was shown that sequencing more tumors could identify additional clinically relevant mutations, the consensus among geneticists is that there should be a shift in focus from sequencing to data analysis (Ledford 2015) by expanding the scope and including clinical data.

5.6   CONCLUSIONS

Reflections on the Human Genome Project and lessons learned therein clearly point to the need to embrace partnerships that span disciplines and geographic boundaries, as well as data sharing (Green et al. 2015). Early design of data analysis plans was identified as a bottleneck that could have made the project more efficient. One of the key take-home messages is to have a bold vision even when there is no clear path toward attaining the goal. It is in this context that we make a case for the NBT approach, which advocates intellectual integration that is discipline-free at the problem statement/definition stage (Huang et al. 2013). Careful consideration of the problem at the early design stage will expedite novel discoveries by integration and analysis of omics data in the context of electronic health records and such personal data. While an immediate path is not clear when we start trying to connect the dots in disparate areas that address the same issue (e.g. human health/disease in the global ecosystem), defining the problem taking into account all aspects by including all stakeholders in the conversation early on will identify the barriers to success of the proposed goal. Application of NBT requires a fundamental change in how we conduct research in academia, how we reward research in academia, and most importantly how we train current and future generations of scientists in NBT. Key to the success of NBT is communication, which is often overlooked in current training programs in biology at all levels. An underlying assumption for working in this exciting data deluge era is open-mindedness and flexibility in researchers. As the scale of data generated is enormous, we can anticipate a shift in hypothesis-driven research toward empirical research as n = all. For example, machine learning algorithms that learn to recognize patterns from empirical data and predict future events and/or make decisions will likely be the foundation for a paradigm shift in health research, from causation to correlation (Liu et al. 2013; Lin and Lane 2017).

In this ever-evolving genomic landscape we envision application of NBT to combine multi-omics data with other types of health data in an iterative process. Grand challenges in biology for societal benefit cannot be resolved in a single conversation. Application of NBT will be a continuous process where scientists from different disciplines (e.g. biology, public health), healthcare providers, practitioners, mathematicians, social scientists, genomicists, and policy-makers get together to address the grand challenges (e.g. health disparities in the United States). Initial interactions will likely focus on overcoming the discipline barriers to speak a common language for sharing ideas. Once there is a common understanding, the team can devise a plan of action and identify additional expertise that needs to be included. Execution of the workflow based on consensus could be altered, if necessary, as additional input becomes available. If we do not include as many stakeholders interested in the outcome of the effort at the problem statement phase, as proposed in NBT, it will not be possible to identify gaps in knowledge that need to be addressed in both the short and long term. The NBT approach will also facilitate identification of the need for novel computational, mathematical, and statistical methods to translate data to wisdom.

Proteomics and transcriptomics research has experienced significant growth in the past 10 years, and the amount of data generated is growing at a remarkable rate. Interdisciplinary approaches have been common in proteomics and transcriptomics research from the beginning. However, technology continues to develop and the data grows further. As precision medicine continues its shift to the forefront of biomedical research, additional expertise is needed from data scientists and healthcare practitioners, among others. An NBT approach is ideal to take advantage of this expertise as the community decides how to approach new research problems and grand challenges in science.

REFERENCES

Aderem A, Adkins JN, Ansong C, et al., 2011. A systems biology approach to infectious disease research: innovating the pathogen–host research paradigm. *MBio*, 2(1):e00325–00310.

Aebersold R, Mann M, 2016. Mass-spectrometric exploration of proteome structure and function. *Nature*, 537(7620):347–355.

Aebersold R, Burlingame AL, Bradshaw RA, 2013. Western blots versus selected reaction monitoring assays: time to turn the tables? *Mol Cell Proteomics*, 12(9):2381–2382.

Anders S, Huber W, 2010. Differential expression analysis for sequence count data. *Genome Biol*, 11(10):R106.

Auton A, Brooks LD, Durbin RM, et al., 2015. A global reference for human genetic variation. *Nature*, 526(7571):68–74.

Barrera NP, Isaacson SC, Zhou M, et al., 2009. Mass spectrometry of membrane transporters reveals subunit stoichiometry and interactions. *Nat Methods*, 6(8):585–587.

Bell AW, Deutsch EW, Au CE, et al., 2009. A HUPO test sample study reveals common problems in mass spectrometry-based proteomics. *Nat Methods*, 6(6):423–430.

Benoist C, Lanier L, Merad M, et al., 2012. Consortium biology in immunology: the perspective from the Immunological Genome Project. *Nat Rev Immunol*, 12(10):734–740.

Bereman MS, Lyndon MM, Dixon RB, Muddiman DC, 2008. Mass measurement accuracy comparisons between a double-focusing magnetic sector and a time-of-flight mass analyzer. *Rapid Commun Mass Spectrom*, 22(10):1563–1566.

Chen R, Mias GI, Li-Pook-Than J, et al., 2012. Personal omics profiling reveals dynamic molecular and medical phenotypes. *Cell*, 148(6):1293–1307.

Dang X, Scotcher J, Wu S, et al., 2014. The first pilot project of the consortium for top-down proteomics: a status report. *Proteomics*, 14(10):1130–1140.

Deutsch EW, Csordas A, Sun Z, et al., 2017. The ProteomeXchange consortium in 2017: supporting the cultural change in proteomics public data deposition. *Nucleic Acids Res*, 45(D1):D1100–D1106.

Ellis MJ, Gillette M, Carr SA, et al., 2013. Connecting genomic alterations to cancer biology with proteomics: the NCI Clinical Proteomic Tumor Analysis Consortium. *Cancer Discov*, 3(10):1108–1112.

ENCODE Consortium, 2012. An integrated encyclopedia of DNA elements in the human genome. *Nature*, 489(7414):57–74.

Grabherr MG, Haas BJ, Yassour M, et al., 2011. Full-length transcriptome assembly from RNA-Seq data without a reference genome. *Nat Biotechnol*, 29(7):644–652.

Green ED, Guyer MS, National Human Genome Research Institute, 2011. Charting a course for genomic medicine from base pairs to bedside. *Nature*, 470 (7333):204–213.

Green ED, Watson JD, Collins FS, 2015. Human Genome Project: twenty-five years of big biology. *Nature*, 526(7571):29–31.

Hood L, Friend SH, 2011. Predictive, personalized, preventive, participatory (P4) cancer medicine. *Nat Rev Clin Oncol*, 8(3):184–187.

Huang X, Bruce B, Buchan A, et al., 2013. No-boundary thinking in bioinformatics research. *BioData Min*, 6(1):19.

International HapMap, 2003. The International HapMap Project. *Nature*, 426 (6968):789–796.

Jay JJ, Eblen JD, Zhang Y, et al., 2012. A systematic comparison of genome-scale clustering algorithms. *BMC Bioinformatics*, 13(Suppl. 10):S7.

Kim MS, Pinto SM, Getnet D, et al., 2014. A draft map of the human proteome. *Nature*, 509(7502):575–581.

Lander ES, Linton LM, Birren B, et al., 2001. Initial sequencing and analysis of the human genome. *Nature*, 409(6822):860–921.

Laney D, 2001. 3D data management: controlling data volume, velocity and variety. META Group Research Note 6.

Ledford H, 2015. End of cancer-genome project prompts rethink. *Nature*, 517 (7533):128–129.

Leinonen R, Sugawara H, Shumway M, International Nucleotide Sequence Database Collaboration, 2011. The sequence read archive. *Nucleic Acids Res*, 39:D19–D21.

Li Z, Zhang Z, Yan P, et al., 2011. RNA-Seq improves annotation of protein-coding genes in the cucumber genome. *BMC Genomics*, 12:540.

Lin E, Lane HY, 2017. Machine learning and systems genomics approaches for multi-omics data. *Biomark Res*, 5:2.

Liu C, Che D, Liu Song Y, 2013. Applications of machine learning in genomics and systems biology. *Comput Math Methods Med*, 2013:587492.

Lockhart DJ, Dong H, Byrne MC, et al., 1996. Expression monitoring by hybridization to high-density oligonucleotide arrays. *Nat Biotechnol*, 14(13):1675–1680.

Lopez-Maestre H, Brinza L, Marchet C, et al., 2016. SNP calling from RNA-seq data without a reference genome: identification, quantification, differential analysis and impact on the protein sequence. *Nucleic Acids Res*, 44(19):e148.

Martens L, Vizcaino JA, 2017. A golden age for working with public proteomics data. *Trends Biochem Sci*, 42(5):333–341.

Morris KV, Mattick JS, 2014. The rise of regulatory RNA. *Nat Rev Genet,* 15(6):423–437.

Nesvizhskii AI, 2014. Proteogenomics: concepts, applications and computational strategies. *Nat Methods,* 11(11):1114–1125.

Pavlou MP, Diamandis EP, Blasutig IM, 2013. The long journey of cancer biomarkers from the bench to the clinic. *Clin Chem,* 59(1):147–157.

Pepe MS, Etzioni R, Feng Z, et al., 2001. Phases of biomarker development for early detection of cancer. *J Natl Cancer Inst,* 93(14):1054–1061.

Perry RH, Cooks RG, Noll RJ, 2008. Orbitrap mass spectrometry: instrumentation, ion motion and applications. *Mass Spectrom Rev,* 27(6):661–699.

Picotti P, Aebersold R, 2012. Selected reaction monitoring-based proteomics: workflows, potential, pitfalls and future directions. *Nat Methods,* 9(6):555–566.

Piskol R, Ramaswami G, Li JB, 2013. Reliable identification of genomic variants from RNA-seq data. *Am J Hum Genet,* 93(4):641–651.

Plebani M, 2005. Proteomics: the next revolution in laboratory medicine? *Clin Chim Acta,* 357(2):113–122.

Poste G, 2011. Bring on the biomarkers. *Nature,* 469(7329):156–157.

Roberts A, Pimentel H, Trapnell C, Pachter L, 2011. Identification of novel transcripts in annotated genomes using RNA-Seq. *Bioinformatics,* 27(17):2325–2329.

Robinson MD, McCarthy DJ, Smyth GK, 2010. edgeR: a bioconductor package for differential expression analysis of digital gene expression data. *Bioinformatics,* 26(1):139–140.

Seyednasrollah F, Laiho A, Elo LL, 2015. Comparison of software packages for detecting differential expression in RNA-seq studies. *Brief Bioinform,* 16(1):59–70.

Smith TF, Waterman MS, 1981. Identification of common molecular subsequences. *J Mol Biol,* 147(1):195–197.

Sowe SK, Zettsu K, 2014. Curating big data made simple: perspectives from scientific communities. *Big Data,* 2(1):23–33.

Tang B, Wang Y, Zhu J, Zhao W, 2015. Web resources for model organism studies. *Genomics Proteomics Bioinformatics,* 13(1):64–68.

Thorisson GA, Smith AV, Krishnan L, Stein LD, 2005. The International HapMap Project web site. *Genome Res,* 15(11):1592–1593.

Toby TK, Fornelli L, Kelleher NL, 2016. Progress in top-down proteomics and the analysis of proteoforms. *Annu Rev Anal Chem (Palo Alto Calif),* 9(1):499–519.

UniProt Consortium, 2017. UniProt: the universal protein knowledgebase. *Nucleic Acids Res,* 45(D1):D158–D169.

Velculescu VE, Zhang L, Vogelstein B, Kinzler KW, 1995. Serial analysis of gene expression. *Science,* 270(5235):484–487.

Vizcaino JA, Csordas A, del-Toro N, et al., 2016. 2016 update of the PRIDE database and its related tools. *Nucleic Acids Res*, 44(D1):D447–D456.

Vorontsov EA, Rensen E, Prangishvili D, Krupovic M, Chamot-Rooke J, 2016. Abundant lysine methylation and N-terminal acetylation in sulfolobus islandicus revealed by bottom-up and top-down proteomics. *Mol Cell Proteomics*, 15(11):3388–3404.

Wehling M, 2021. *Principles of Translational Science in Medicine: From Bench to Bedside.*. Amsterdam: Elsevier.

Wilhelm M, Schlegl J, Hahne H, et al., 2014. Mass-spectrometry-based draft of the human proteome. *Nature*, 509(7502):582–587.

Wolfe CJ, Kohane IS, Butte AJ, 2005. Systematic survey reveals general applicability of "guilt-by-association" within gene coexpression networks. *BMC Bioinformatics*, 6:227.

Xie Y, Wu G, Tang J, et al., 2014. SOAPdenovo-Trans: de novo transcriptome assembly with short RNA-Seq reads. *Bioinformatics*, 30(12):1660–1666.

Zhang Y, Fonslow BR, Shan B, Baek MC, Yates JR, 2013. Protein analysis by shotgun/bottom-up proteomics. *Chem Rev*, 113(4):2343–2394.

Zubarev RA, 2013. The challenge of the proteome dynamic range and its implications for in-depth proteomics. *Proteomics*, 13(5):723–726.

6 Pharmacogenomics

Alison Motsinger-Reif

6.1 INTRODUCTION

Pharmacogenomics is the study of genetic factors that influence the variability in an individual's or a population's response to a drug. The ultimate goal of pharmacogenomics is to use an individual's or a population's genomic information to optimize clinical treatment, prevent adverse reactions, and improve patient care. Pharmacogenomics falls within the broader area of personalized medicine or precision medicine. In 2015, the Precision Medicine Initiative (PMI) was launched, involving the National Institutes of Health and several government and academic institutions. The PMI Working Group defines precision medicine as "an approach to disease treatment and prevention that seeks to maximize effectiveness by taking into account individual variability in genes, environment, and lifestyle" (PMI Working Group 2015). It is clear that pharmacogenomics is a key discipline in this important initiative.

Pharmacogenomics has the potential to change the way drugs are prescribed. The outcomes from genomic analyses can be used to identify biomarkers for early diagnosis, optimize clinical treatment, prevent adverse reactions, improve patient care, and reduce healthcare costs. Additionally, pharmacogenomics can help understand the mechanism of action of drugs and may aid in drug development, among other goals. In this chapter we present an overview of the current state and applications of pharmacogenomics, recent research and current methods used in the field, available pharmacogenomics resources, use of pharmacogenomics in clinical practice, and its future applications, along with a discussion of social, ethical, and economic issues.

The history of pharmacogenetics can purportedly be traced as far back as 510 BCE, when the Greek philosopher and mathematician Pythagoras reported that certain individuals experienced severe abdominal pain and cramps upon eating fava beans. This was later validated and identified as hemolytic anemia, which is caused by glucose-6-phosphate dehydrogenase (G6PD) deficiency (Nebert 1999). The term pharmacogenetics was coined by Vogel in 1957 (Vogel 1959). Pharmacogenetics is the study of a single gene or DNA variant that may be associated with a dose or drug response phenotype. Pharmacogenomics is the study of the entire genome to find variants that may be associated with dose–response phenotypes. There have been a number of technological advances that have facilitated the transition from pharmacogenetics to pharmacogenomics: (1) the completion of various human genome sequencing and mapping projects, such as the Human Genome Project (International Human Genome Sequencing Consortium 2001), the International HapMap project (International HapMap Consortium 2003), and the 1000 Genomes Project (1000 Genomes Project Consortium 2015); (2) recent technological and computational advancements in high-throughput genotyping and sequencing; and (3) the evolution of statistical analyses methods and tools such as association mapping and pathway analyses.

Research has consistently shown that the dose–response phenotypes across a number of drugs are influenced by several genetic and nongenetic factors. The PharmGKB project actually maintains a list of drugs that have pharmacogenetic information on the official drug label (www.pharmgkb.org/view/drug-labels.do). As of December 2016 there were 235 drugs with pharmacogenetic information on the label (PharmGKB 2017a). The variability in response to a drug is observed not only between individuals, but also between populations (Bachtiar and Lee 2013). Dose response is now appreciated as a complex trait that is influenced by a number of genetic, epigenetic, and environmental factors and their interactions, and mapping techniques and the corresponding analysis tools available have advanced as technology has progressed (Weigelt and Reis-Filho 2014; Jones et al. 2016).

What is meant by a drug response phenotype can vary between studies. Phenotypes can relate to any aspect of drug response, such as efficacy, metabolism, absorption, dosing requirements, or toxicity. Pharmacogenomic data could be used to select the most efficacious drug for a patient based on their ancestry or population subgroup. Dose selection can be done much more precisely based on the patient's genotype to determine the dosage required to achieve the optimal drug effect for that patient. Adverse reactions to drugs can be proactively prevented by cross-checking pharmaco-genomic biomarkers for gene–drug pairs. This is the aim of precision medicine – to prescribe medications to an individual or a population based on their genetic make-up, thus leading to optimized clinical treatments and minimized adverse reactions and reduced healthcare costs.

In this chapter we cover the current state of the field of pharmacogenomics, providing high-impact example success stories. We briefly discuss some of the major study designs and genotyping resources available to pharmacogenomics studies, while trying to focus on different advantages and disadvantages of the resources that are specific to pharmacogenomics. Additionally, we highlight open challenges in the area, with specific emphasis on hurdles to implementation of successful associations.

6.2 THE CURRENT STATE

Over the last two decades there has been an increase in the number of genetic variations found to be associated with variability in drug response. As mentioned above, there are over 200 drugs with genes listed on the drug label, and a large number of associations that have been detected and replicated, but have not yet progressed within the translation process to inclusion on the label (Shuldiner et al. 2013). Later in the chapter we will highlight some specific successes in the field. In this section we will discuss the tools and methods available for discovering such associations.

6.2.1 Methods Used in Pharmacogenomics

In this section we will discuss the various study designs and methods for molecular analysis, quality control, and statistical analysis used in pharmacogenomics. These methods have been reviewed by Ritchie (2012), and we provide additional depth here.

6.2.1.1 Study Designs

The main study designs available in pharmacogenomics are *in vitro* studies, randomized clinical trials, case–control studies, and biobanks linked to electronic medical records (Ritchie 2012). We briefly discuss each of these study designs along with their advantages and disadvantages. Pharmacogenomics mapping can theoretically be performed in any study design (long-term cohort, family-based linkage studies, etc.), but there are practical issues that restrict the commonly used study designs. For example, family-based studies are the gold standard for establishing the genetic component of complex traits, but are rarely available for drug response studies. Practicalities of collecting data on drug response across multiple generations of a family, all taking the medication for the same indication, make this study design extremely rare. Because of this, the most commonly used study designs are *in vitro* models, clinical trials, case–control studies, and electronic medical record cohorts. It is important to note that this results in a crucial assumption in pharmacogenomics, that may or may not always be true. Without family-based designs there is no way to estimate the heritability of drug response phenotypes – the variation in the phenotype that is explained by genetic variation. This is routinely estimated for complex traits, and is a necessary yet not sufficient step in gene mapping. It is possible that a trait can have a genetic etiology but very low heritability. Heritability is essential for association mapping to be successful. Without family-based designs in pharmacogenomics, the assumption that traits are heritable goes untested (Motsinger-Reif et al. 2013).

In vitro model systems are a powerful approach for drug response gene mapping. While there are a number of cell lines available, lymphoblastoid cell lines (LCLs) are commonly used (Jack et al. 2014). Human lymphocytes are infected with the Epstein–Barr virus (EBV) to create immortalized human LCLs. These LCLs can be easily maintained in the lab and can serve as a constant source of cell lines for dose–response assays. The use of LCLs in pharmacogenomic studies has several advantages. First, there is a variety of study designs available with the LCLs. Both unrelated and pedigree-bases resources are available. This pedigree data allow for the estimation of heritability in drug response traits, which as mentioned above is rarely available in patient data. Additionally, LCLs are available from public resources such as the CEPH pedigrees (Dausset et al. 1990), the International HapMap project (International HapMap Consortium 2003), and the 1000 Genomes Project (1000 Genomes Project Consortium, 2015), along with publicly available genome-wide single-nucleotide polymorphism (SNP) data. Gene expression data are also available for cell lines from the 1000 Genomes Project. This removes the cost of genotyping for association mapping studies with this model. The LCL model is conducive to robotic automation for high-throughput dose–response assays, allowing for a large number of drugs and cell lines to be tested. Thus, the LCL model is a highly scalable, efficient, and cost-effective system compared to other models used in pharmacogenomics. Compared to other pharmacogenomic study designs such as clinical trials, the LCL model has fewer limitations with respect to sample size, confounders such as comorbidities, multiple-treatment regimens, and uncontrolled environmental factors. Potential confounders such as LCL growth rate and batch effects can be controlled for by careful and methodical technical execution and quality control of the data (Choy et al. 2008). The use of LCLs has been shown to be a successful model for drug response assays in pharmacogenomics (Wheeler and Dolan 2012; Jack et al. 2014). The LCL model has had several successes in pharmacogenomics, such as SNP associations in the *MGMT* gene with temozolomide response (Brown et al. 2012a), dose–response characterization of 29 FDA-approved drugs (Peters et al. 2011), identification of loci associated with camptothecin-induced cytotoxicity (Watson et al. 2011), and many others.

Of course, the cell line model does have disadvantages. One key disadvantage is that cell lines can only be made from certain tissues, which may or may not be the same as the target tissue of the drug. For example, LCLs which are developed from lymphocytes are used to assay drug response for multiple types of cancer such as breast cancer, ovarian cancer, lung cancer, etc. Cell lines may not express the enzymes involved in drug absorption, distribution, metabolism, and excretion (ADME) which are the important endophenotypes for the pharmacokinetics and pharmacodynamics of the drug. The process of immortalization of the cell lines may alter cellular characteristics that may affect the response to the drug. In spite of these drawbacks, the cell line model has proven to be very informative and successful in pharmacogenomics (Welsh et al. 2009; Zilak et al. 2011; Cox et al. 2012; Wheeler and Dolan 2012).

A randomized clinical trial (RCT) is the benchmark for drug development studies, and is one of the most commonly used study design types in pharmacogenomics. The main goal of a clinical trial is to determine the efficacy and toxicity of the new drug. An RCT usually consists of multiple-treatment arms and sometimes also includes a control arm. In order to minimize patient recall bias and researcher bias, RCTs are typically conducted in a randomized and double-blind manner. Patients are randomly assigned to a treatment arm and the researcher collecting the data is unaware of which treatment arm the patient is assigned to. Patient demographics such as sex, age, race, ethnicity, current physiological conditions, other medications, diet, and various environmental factors can be recorded during the trial. The patient's response to the drug, toxicity assessment, side-effects, and various endophenotypes such as drug clearance, metabolite levels, etc. can be measured from the patients. Various statistical analyses methods can be employed to determine the association between the drug response phenotypes and the patient genotypes, gene × environment (here, drug) interaction and various other variables. The RCTs themselves can be guided using genomic data to reduce the heterogeneity among patients enrolled in the trial. For example, if a DNA variant is known to influence the response of the drug under development, say from previous *in vitro* studies or biochemical knowledge of the drug, then the RCT can be performed by enrolling only those patients carrying the specific DNA variant. This would increase the power of the RCT, making it much more efficient and thus reduce the cost of drug development (Harper and Topol 2012).

RCTs have several advantages for pharmacogenomics. Unlike *in vitro* model systems, the dose–response traits are measured directly in humans and are likely to be immediately relevant. Drug response phenotypes, including ADME endophenotypes, can be directly measured from patients enrolled in the trial to determine the impact of the drug on the patients. Unfortunately, there are also several limitations with RCTs for gene mapping. They are expensive to conduct, with very large trials costing hundreds of millions of dollars. The design of the trial is typically focused on the efficacy questions; they are rarely designed for gene mapping. Because of this, sample sizes are usually not sufficient for gene mapping, leading to low statistical power. Patients enrolled in the trial may leave, die, or be forced to leave due to complications, new comorbidities, or new drug regimens, thus leading to incomplete cases and further reduction in sample size. There may be unknown confounders or uncontrolled parameters such as diet or other environmental factors. Also, results from the RCT are only valid for the population enrolled in the RCT. Enrolling large number of individuals of different races, ethnicities, and other demographics is usually untenable and increases the cost of the trial (Ritchie 2012; Stolberg et al. 2004).

An alternative study design available for pharmacogenomics is a retrospective study. Both case–control and observational studies are retrospective in nature and are usually less expensive than RCTs. In pharmacogenomic case–control studies all individuals enrolled receive the drug treatment. Individuals are classified into the case or the control group based on the drug response phenotype – for example, responders to treatment vs. nonresponders to treatment, or individuals who had an adverse reaction to the drug vs. those who did not. Demographic and genomic information about the patients is collected retrospectively. This is a widely used and less expensive alternative to an RCT and sample sizes can be controlled for the cases and controls. It is important that the cases and controls are drawn from the same population so that any associations found are truly due to effect of genotype and not due to population stratification. This type of study suffers from several limitations, such as patient recall bias, uncontrolled factors such as diet, environment, multiple drug regimens, changes in drug dosage, and more (Kraft and Cox 2008; Ritchie 2012).

The use of electronic medical records (EMRs) or electronic health records (EHRs) is gaining popularity in pharmacogenomic studies. Electronic health records provide a rich source of clinical data that include patient demographics, biological and environmental factors, patient medical history, physiological conditions, and health outcomes. The implementation of EHRs in a majority of hospitals and healthcare systems in the United States allows for the feasible collection and access of these records. The data for the one million volunteers of the PMI cohort will be collected and shared using EHRs in biobanks (PMI Working Group 2015).

The Electronic Medical Records and Genomics Network (eMERGE) was established by the National Human Genome Research Institute (NHGRI) to explore the utility of genomic data in EMRs (McCarty et al. 2011; Crawford et al. 2014). The eMERGE-PGx project, a partnership of eMERGE and the Pharmacogenomics Research Network (PGRN), has deployed a multi-site EHR database consisting of DNA variant information in proposed pharmacogenes in ~9000 patients under drug treatment, with the goal of developing a database of pharmacogenetics variants linked to drug phenotypes to inform precision medicine in the clinic and provide data for pharmacogenomic studies (Rasmussen-Torvik et al. 2014). The use of EHRs has led to a number of successful pharmacogenomic discoveries.

Dahlin et al. (2015) identified genetic variants associated with response to inhaled corticosteroids in asthma patients using EHRs from BioVu (Roden et al. 2008) at Vanderbilt University Medical Center and the Personalized Medicine Research Project (PMRP) (Wilke et al. 2007) at the Marshfield Clinic. Low et al. carried out genome-wide association studies (GWAS) using data from BioBank Japan to identify genetic variants associated with chemotherapeutic drug-induced toxicity in 13,122 cancer patients (Low et al. 2013). DNA samples from EHRs from BioVu (Roden et al. 2008) were used to predict drug response in patients on clopidogrel treatment, supporting the use of EHRs for pharmacogenomic studies (Delaney et al. 2012). The wide implementation and availability of EHRs and the merging of data from multiple sites into pooled EHRs has led to an increase in the number of records in the database, and association studies can be developed from these populations.

However, there are limitations to this study design with respect to the availability of samples with the phenotype of interest or patients treated with the drug under study. Not having sufficient samples limits the power of analyses conducted from EHRs. The data extant in the biobank may not be sufficient for the research being conducted. Sophisticated informatics, data mining, and analytics are required to extract the data required for the research. Regardless of these challenges, EHRs offer an enormous opportunity for pharmacogenomic discoveries.

6.2.1.2 Molecular Methods

The molecular assays available for pharmacogenomic studies are similar to those for other genetic and genomic studies. There are both candidate gene and genome-wide options in pharmacogenomics, and data can be collected with any available genotyping technology, from SNP chips to next-generation sequencing, etc. The choice of genotyping assay depends on a variety of factors.

Candidate gene studies can be performed using a variety of assays, from targeted genotyping of individual polymorphisms, to specifically designed SNP chips for pharmacogenomics. Because there is often a lot of information available about the mechanism of action of the drugs studied in pharmacogenomics, this information can be used to choose candidate genes. In fact, there are candidate gene platforms developed specifically for pharmacogenomics. Both Affymetrix and Illumina offer pharmacogenetic candidate gene chips (Burmester et al. 2010; Illumina 2017). They include approximately 2000 variants in genes that include the cytochrome P450s (CYP450s), the key metabolizing enzymes, many other enzymes involved in phase I and phase II pharmacokinetic reactions, and signaling mediators associated with variability in clinical response to numerous drugs not only among individuals, but also between ethnic populations. Generally, candidate gene selection and variant selection can be performed just as it would be in any gene mapping study. A recent article by Patnala et al. (2013) reviews approaches and tools available for this selection. Because there is often a generous amount of information on the mechanism of action of many drugs, such information can be used to help guide candidate gene selection.

This information on mechanism can also help guide choice of molecular assay. For example, if the drug target is in a genic region, then exome sequencing may be appropriate and will reduce the number of variants by several magnitudes as compared to whole-genome sequencing. The selection of genotyping platform also depends on whether the study being conducted is a candidate gene study or a whole-genome study. If the drug target is specific and well-known, a candidate gene study is appropriate and sufficient and specific sequencing arrays can be used. For genome-wide studies, commercially available genotyping platforms are used to capture the common SNPs in various populations. Next-generation sequencing (NGS) can also be used. For example, the eMERGE-PGx project has launched PGRNSeq, which is an NGS platform that captures DNA variation in target pharmacogenes (Rasmussen-Torvik et al. 2014).

Another crucial aspect of association mapping is phenotype definition. For most complex traits in human genetics, phenotype definition is a biological or clinical question, and pharmacogenomics phenotypes certainly need to be defined in these contexts as well. It is notable, however, that in pharmacogenomics the phenotype itself is often a molecular measurement, or the result of modeling. Intermediate phenotypes such as metabolite levels could also be measured. Several phenotypes such as efficacy, toxicity levels, and other phenotypes that involve the ADME features of the drug can also be modeled (Welsh et al. 2009).

6.2.1.3   Quality Control Methods

As with any study, quality control (QC) of the data prior to statistical analyses is crucial. All QC measures that apply to genetic studies apply to pharmacogenomic studies as well. Potential confounders, batch effects, outlier detection, and biases such as population stratification must be corrected or controlled for before any statistical analyses are performed. The genotype data must also be filtered for missing data, SNPs in high linkage disequilibrium, deviation from expected proportions under Hardy–Weinberg equilibrium, and low minor allele frequencies. If individuals in the study are from different populations, then their allele frequencies will differ and this population substructure must be controlled for. The most common method for controlling population substructure is to use principal components analysis (Ringnér and Ringner 2008). Quality control methods for genetic studies have been discussed in detail and recommendations and guidelines have been provided by Jack et al. (2014), Choy et al. (2008), Laurie et al. (2010), and Zuvich et al. (2011).

While the majority of the QC procedures are identical between pharmacogenomics studies and other association mapping studies, there are a couple of unique areas to highlight. First, when testing for deviations from Hardy–Weinberg proportions, most software packages assume that variants are biallelic. A large number of variants for common candidate genes in pharmacogenomics are not biallelic. Special consideration needs to be taken for such variants in QC. Additionally, because many of the phenotypes used in pharmacogenomics are actually molecular measurements, or summarizations from initial modeling steps, there is often considerable data cleaning and modeling of the phenotype variables.

6.2.1.4   Statistical Methods

A variety of statistical methods are used in pharmacogenomics analyses. The choice of the statistical method used is guided by the study design and the hypothesis under test. Standard methods – such as simple and multiple linear regression, logistic regression for binary responses, and chi-square tests – are widely applicable (Ritchie 2012). Repeated measures analysis can be used for study designs in which the same patients are treated with different doses of a drug repeatedly over a period of time. For a candidate gene or SNP study, an analysis of variance (ANOVA) method is appropriate to examine the differences in the drug response phenotype for the variants under study.

When multiple factors or variables are modeled, variable selection or model selection methods such as ranking methods or sequential methods can be used to build parsimonious models. Ranking methods build models with all possible combinations of variables and pick the top-ranking method based on relative ranking criteria such as the adjusted R^2, predicted sum of squares (PRESS) statistic, Akaike information criterion (AIC) (Akaike 1974), Bayesian information criterion (BIC) (Schwarz 1978), etc. Sequential methods use forward selection, backward elimination, or stepwise selection to iteratively build the model and include or exclude variables based on relative performance of the compared models measured using the adjusted R^2, PRESS statistic, AIC, BIC, etc. Ranking methods have better statistical properties over sequential methods and hence are preferred, but can be computationally intensive if the number of variables is very large.

When multiple hypothesis tests are performed, multiple testing correction must be done to control the Type I error rate. The simplest and most widely used multiple testing correction method is the Bonferroni correction. If the experiment-wise Type I error rate is set to α, and m independent tests are performed, then the Bonferroni corrected significance threshold for each test is set to α/m. The Bonferroni correction is very conservative, especially when the individual tests are not independent. In such cases, a false discovery rate (FDR) method may be more appropriate (Benjamini and Hochberg 1995). The FDR method adjusts p-values to control the number of false positive results from the positive results instead of from all results. The adjusted p-values are called q-values. The FDR method has increased power over the Bonferroni correction but could lead to increased Type I error rates. The multiple testing correction method and the significance threshold chosen depend on the cost of Type I and Type II error rates. If the tests are hypothesis-generating, then a less stringent FDR method may be appropriate, while more strict methods may be appropriate for follow-up or hypothesis-replicating tests (Noble 2009).

For genome-wide studies of SNPs, the statistical methods used to test for association are typically identical to those used in candidate gene studies. Typically, variable selection is performed prior to entering any SNP information into the model using the methods mentioned above. Then, a term including an individual SNP is entered into the model for each SNP in the dataset. Due to the enormously large number of tests done in a GWAS, multiple testing correction must be done to reduce the Type I error rate. The standard threshold for significance in a GWAS, using Bonferroni correction, is $p < 5 \times 10^{-8}$. This represents a Bonferroni correction for the effective number of tests across the genome (Li et al. 2012). For studies with smaller sample sizes this multiple testing correction may be too conservative; in these cases suggestive levels of significance ($p < 5 \times 10^{-6}$ or $p < 5 \times 10^{-5}$) are often considered. Various other methods, such as permutation testing methods (Che et al. 2014), SLIP, and SLIDE (Han et al. 2009) have been applied to GWAS data. A permutation test is used when the distribution under the null is not known. A permutation test or a randomization test is used to permute the observed data by randomly assigning the phenotype to each observation to obtain the sampling distribution under the null hypothesis, which is then compared to the distribution obtained from the actual observed data. A permutation test is a nonparametric test but is computationally intensive and works best for case–control studies. An assessment of different multiple comparisons methods is presented by Johnson et al. (2010).

There has been a steady increase in the number of reported GWAS as per the GWAS Catalog of NHGRI-EBI (Burdett et al. 2017). The tools, software, interpretation, and analyses of GWAS are rapidly evolving with its prevalent use. When results from multiple GWAS with comparable test statistics for examining the same phenotype are available, these results can be combined using meta-analysis to derive a pooled estimate. The meta-analysis approach is extensively used to increase power and to prioritize GWAS results (Yesupriya et al. 2008; Cantor et al. 2010). There are several challenges to performing a meta-analysis, such as heterogeneity in study design, test hypotheses, phenotype measurement, and summary test statistics. Careful planning and cooperation between the participating groups is required to prevent these disparities. A genome-wide meta-analysis that combined GWAS results from seven previously published HapMap panels for carboplatin- and cisplatin-induced toxicity in LCLs from a diverse population identified SNPs and genes associated with platinum-induced cytotoxicity, thus showing the success of the meta-analysis approach (Wheeler et al. 2013). Various software packages such as METAL (Willer et al. 2010), metaan (Stata) (Kontopantelis and Reeves 2010), and Comprehensive Meta-Analysis (Meta-Analysis 2017) are available to perform meta-analysis of large datasets.

To support associations found from statistical analyses, such as GWAS, a "Triangle model or approach" is commonly used. The first stage in this model identifies the genomic markers (e.g. SNPs) associated with the phenotype. The second stage filters the identified SNPs to find the SNPs that are expression quantitative trait loci (eQTLs) – that is, the SNPs that influence the expression level of one or more genes. The final stage attempts to find associations of the genes identified in the second stage with the phenotype (Welsh et al. 2009; Jack et al. 2014). This approach provides stronger evidence of the SNP–phenotype association by elucidating the mechanism of action for the association and also because SNPs associated with phenotypic outcomes are more likely to be eQTLs (Huang et al. 2007; Nicolae et al. 2010).

The statistical tests mentioned so far test for the association of single or multiple factors with a single response. To model multiple factors or variables with multiple response variables, multivariate analysis of variance (MANOVA) or multivariate analysis of covariance (MANCOVA) methods are commonly used. These methods can incorporate multivariate phenotypes (such as an entire dose–response profile instead of a univariate dose–response summary). These multivariate methods have been shown to be more powerful than the methods using the univariate response (Brown et al. 2012b, 2014; Jack et al. 2014). The methods listed above are available in various software packages such as R (R Development Core Team 2016), SAS® (SAS 2017), and PLINK (Purcell et al. 2007).

While a GWAS tries to find SNPs that are associated with a given phenotype, in contrast, a phenome-wide association study (PheWAS) tries to find multiple phenotypes that are associated with a given SNP. Results from PheWAS have successfully identified SNP–phenotype associations and have revealed potential pleiotropic effects of genetic markers (Denny et al. 2010, 2011, 2013). PheWAS can also be used for pharmacogenomic studies, as demonstrated by Moore et al. (2015), who use PheWAS on human immunodeficiency virus (HIV) clinical trials data to identify associations with clinical phenotypes. PheWAS has also been used to identify clinical phenotypes associated with thiopurine S-methyltransferase (TPMT) enzyme activity level in patients treated with thiopurine drugs (Neuraz et al. 2013).

The models so far have only considered the influence of a single genetic marker on the dose–response phenotype. However, the dose–response phenotypes or pharmacogenomic traits are complex traits influenced by multiple genomic markers and their interactions with each other and with the environment (Moore 2003). For example, genetic variations associated with drug response in cancer treatment have been found to be context-dependent, and drug resistance in cancer tumors is believed to be due to the epistatic interactions of a wide array of constantly evolving genetic aberrations in tumor cells (Weigelt and Reis-Filho 2014). Various data mining tools and methods can be used to detect epistasis in pharmacogenomic studies and have been reviewed by Motsinger et al. (2007). Classification and regression trees can be used to interpret the relationship between the genetic and environmental factors affecting the dose–response phenotype and/or to predict the response. Multi-tree methods such as random forests (RF) and bagging and boosting build a large number of trees and use a voting or consensus method across all trees to predict the response. These tree-based methods can also be used for variable selection.

Random forest can also be used to rank the importance of the variables or factors. The RF method has been shown to have more power for detecting interactions as compared to standard regression methods (Motsinger et al. 2007). Another method to detect high-order gene–gene and gene–environment interactions is the multifactor-dimensionality reduction method (MDR), which uses a form of per-mutation testing and has power even in small-sized samples. MDR reduces the dimensionality of predictor variables by pooling them into high-risk and low-risk groups so that high-order interactions can be modeled even for small numbers of samples, as usually seen in case–control or sib-pair studies (Ritchie and Motsinger 2005).

Neural networks can also be used to model epistatic pharmaco-genomic traits. Neural networks are a computational approach that mimics the information processing of a biological nervous system by learning from a training dataset and making predictions on previously unobserved data. Neural networks are nonparametric models and can handle big data. Genetic programming neural network (Ritchie et al. 2003) and grammatical evolution neural network (Motsinger et al. 2006) are evolutionary computation algorithms that include variable selection and cross-validation in a neural network analysis to find the optimal neural network for the dataset being modeled, and have high power to detect genetic epistasis.

With the increasing scale of data and the availability of "multi-omic" data, pathway and network analysis methods that integrate big data, multi-omic data, and domain knowledge have become popular. Multiple genes or proteins interact within a pathway and multiple pathways interact within a network. The goal of pathway or network analysis is to find all the genetic variants involved in the biological pathway of interest. The gene sets involved in specific biological pathways can be obtained from pathway databases such as Gene Ontology (Ashburner et al. 2000), the Kyoto Encyclopedia of Genes and Genomes (KEGG) (Kanehisa and Goto 2000), the Molecular Signatures Database (MSigDB) (Subramanian et al. 2005), the Network Data Exchange (NDEx) (Pratt et al. 2015), the Pharmacogenomics Knowledgebase (PharmGKB) (Whirl-Carrillo et al. 2012), and many others. Pathway analysis has the advantage of being knowledge-driven and provides insight into the mechanism of action of the genes associated with the phenotype of interest. Khatri et al. (2012) describe available pathway analysis methods and their limitations in detail. For complex networks, various visualization methods and techniques such as graphs, networks, clusters, and heat maps can be used to represent intricate biological networks and to interpret genetic interactions (Merico et al. 2009).

6.2.2 Advantages in Pharmacogenomics

Pharmacogenomic studies have several specific advantages over genomic studies for other complex traits or diseases. The variants in genes associated with pharmacogenomic traits typically have larger effect sizes than those for other complex traits or diseases (Ritchie 2012). This is due to the strong influence of drug exposure, which is absent or unknown in other complex traits. Due to this larger effect size, pharmacogenomic association studies can be conducted with a smaller sample size (Ritchie 2012). While candidate gene studies use the knowledge of the mechanisms involved in the ADME of the drug to identify genetic variants influencing dose response, they cannot be used to identify variants not involved in the known mechanisms. In contrast, the unbiased nature of a GWAS allows for the interrogation of the entire genome and does not have a prerequisite of pharmacologic knowledge. Thus, a GWAS is able to identify novel genetic variants involved in dose response. Advances in genotyping technology now provide more complete coverage of the genome, ensuring that genetic markers influencing dose–response phenotypes are not missed (Motsinger-Reif et al. 2013).

Another crucial advantage of pharmacogenomics is the translational potential. Pharmacogenomic associations can have a relatively quick and direct impact on human health in terms of efficacy and safety of the drugs, and in terms of healthcare costs. Hence, there are strong incentives to rapidly translate the findings from pharmacogenomic studies to drug discovery and development and clinical practice (Harper and Topol 2012).

6.2.3 Challenges in Pharmacogenomics

While there has been an increase in the number of reported GWAS, of the 2785 GWAS listed in the GWAS Catalog, only 183 GWAS are reported for SNP associations with drug response phenotypes, which is only 6.5 percent of the entire reported GWAS (Burdett et al. 2017). Considering that pharmacogenomics is an important and clinically actionable trait, this number is lower than expected. There are several challenges limiting pharmacogenomic studies. Similar to other GWAS, obtaining large sample sizes to provide sufficient statistical power is a challenge in pharmacogenomic GWAS. This problem is exacerbated in pharmacogenomic studies due to the difficulty of finding patients with a particular disease on a particular drug of the many drugs available for the treatment of the disease. This problem becomes even worse when trying to obtain patients with a rare or adverse drug reaction, as these are a subset of the population on a particular drug. Finding patients from different ethnic or minority groups is also a challenge in pharmacogenomic studies. A large number of GWAS are performed on individuals of European ancestry. However, both allelic and genetic heterogeneity across populations can result in unique associations in different ethnic groups. For example, SNPs in the thiopurine methyltransferase (*TPMT*) gene are associated with thiopurine toxicity in individuals of European ancestry, but in individuals of Asian ancestry SNPs in the *NUDT15* gene were found to be more critical (Giacomini et al. 2016).

Validation of results from pharmacogenomic studies requires replication in independent cohorts. With the already existing challenges of finding a suitable discovery cohort, finding appropriately characterized replication cohorts is even more exigent than in other complex traits.

Several approaches have been taken to overcome the challenges faced due to small sample sizes. Various consortia (discussed later in this chapter) have been formed to facilitate the sharing of samples or cohorts from multiple studies (Motsinger-Reif et al. 2013; Giacomini et al. 2016). Meta-analysis approaches are being used to combine results from multiple GWAS to increase power and validate association findings (Ioannidis et al. 2001; Wheeler et al. 2013). Since pharmacogenomic studies often have the advantage of known mechanisms of action for the drug under study, functional validation studies may be more feasible than performing a replication study. For example, siRNA knockdowns of associated genes can be performed to modulate expression levels of drug targets. If the pharmacological target modulates gene expression levels, then RNA-Seq or ChIP-Seq may be performed to identify differentially expressed genes and regulatory elements (Yee et al. 2016).

In addition to sample size limitations, the unavailability of family-based designs, the need for control and treated samples, population stratification, comorbidities, and other environmental confounders are also limitations for pharmacogenomic studies (Motsinger-Reif et al. 2013).

6.3   CURRENT PRACTICE: PHARMACOGENETICS IN THE CLINIC

Several pharmacogenetic discoveries have been successfully translated into clinical practice. Several academic and government organizations provide guidelines to physicians for the use of genetic information to better inform drug prescribing decisions. The Clinical Pharmacogenetics Implementation Consortium (CPIC, https://cpicpgx.org) (Relling and Klein 2011), which is a partnership between the Pharmacogenomics Research Network (PGRN 2017) and the Pharmacogenomics Knowledge Base (Whirl-Carrillo et al. 2012), provides a list of gene–drug pairs along with prescribing guidelines for clinical decision support (CDS). The US Food and Drug Administration (FDA) also lists pharmacogenomic biomarkers for several FDA-approved drugs and requires this information in drug labeling (US FDA 2017). These guidelines are aimed at providing physicians with a curated and peer-reviewed list of gene–drug pairs with specific and actionable recommendations for prescribing drugs based on pharmacogenetic test results.

6.3.1   Clinically Available Pharmacogenetic Tests

The current use of pharmacogenetics in the clinic falls into three major categories: (1) efficacy or therapy selection, (2) dose selection, and (3) risk assessment for adverse drug reaction (ADR). Listed below are some examples of the gene–drug pairs classified into these categories that are listed in the CPIC guidelines and the FDA's pharmacogenomic biomarkers in drug labeling list.

1. Efficacy or therapy selection: Genetic testing can help identify if a patient has a mutation that cannot activate a commonly prescribed medicine. In this case, the patient can be prescribed an alternative therapy. For example, clopidogrel is an inactive prodrug given to patients who have had a heart attack, stroke or have blockages in their blood vessels due to cholesterol accumulation. Clopidogrel is activated *in vivo* by several cytochrome P450 enzymes, some of which are encoded by *CYP2C19*. Variants in *CYP2C19*

can reduce enzymatic function, lowering conversion of the clopidogrel prodrug to its active form, thus reducing its efficacy and increasing the risk of heart attack and stroke (Mega et al. 2009). The CPIC dosing guideline recommends that patients with reduced-function *CYP2C19* variants be prescribed a different antiplatelet drug such as prasugrel or ticagrelor (Scott et al. 2013).

2. Dose selection: Warfarin, an oral anticoagulant, is a commonly prescribed drug to prevent heart attacks, strokes, and blood clots. Genetic polymorphisms in two genes – cytochrome P450, family 2, subfamily C, polypeptide 9 (*CYP2C9*), and vitamin K epoxide reductase complex, subunit 1 (*VKORC1*) – influence the variability in required warfarin dosage among patients (Takeuchi et al. 2009). These polymorphisms influence both the pharmacokinetics and pharmacodynamics variation of warfarin. Variations in *CYP2C9* and *VKORC1* affect the variation in dose response and the risk of the adverse event of bleeding in patients taking warfarin. Previously warfarin dosage was determined empirically on a per patient basis. Current recommendations to physicians are for genotype-guided warfarin dosage calculations (Johnson et al. 2017), which have been shown to increase the efficacy of the drug and reduce the risk of adverse events (Caraco et al. 2008; International Warfarin Pharmacogenetics Consortium 2009; Epstein et al. 2010).

3. Risk assessment for ADR: The human leukocyte antigen (*HLA*) genes regulate the immune system in humans and the proteins encoded by these genes bind to the active compounds in various drugs. *HLA* genes are highly polymorphic and *HLA* gene variants occur at different rates in different populations. HIV patients with *HLA-B*5701* taking the drug abacavir are at risk of developing a life-threatening skin condition known as toxic epidermal necrosis (TEN) (Hetherington et al. 2002). Drug toxicities are the most common reason that HIV patients discontinue treatment. Carbamazepine, which is prescribed for neurological disorders, is another drug that causes fatal skin hypersensitivity through TEN in individuals carrying the *HLA-B*1502* variant (Chen et al. 2011). Pharmacogenomic analyses of the patient's genome can help identify the right drug regimen while minimizing adverse side-effects. The CPIC guidelines recommend *HLA-B* typing before prescribing abacavir or carbamazepine to identify at-risk patients who can either be prescribed very low doses of the drugs or an alternative therapy (Leckband et al. 2013; Martin et al. 2014).

The examples listed above are for germline sequence variations. Pharmacogenetic tests for somatic mutations in tumors are already widely used in all the above categories and are based on specific genetic aberrations in the cancer cells. For example, the HER2 (human epidermal growth factor receptor 2) protein is overexpressed in some cases of breast cancers and gastric cancers. Cancer immunotherapy in these cases involves the use of the drug trastuzumab, a monoclonal antibody, that targets the HER2 protein and has been shown to have improved outcomes over other treatments (Vogel et al. 2002; Marty et al. 2005). Both the CPIC and FDA guidelines require that genetic test results in patients with metastatic breast or gastric cancer indicate HER2 overexpression in tumor cells before prescribing trastuzumab therapy (Pharmacogenomics at work 1998). Several other pharmacogenetics tests for targeted cancer therapy are used in clinical practice, such as the use of the drugs gefitinib or erlotinib in lung cancer patients whose tumors have mutations in the tyrosine kinase domain of the epidermal growth factor receptor (*EGFR*) gene (Pao et al. 2004), or testing for the *BCR-ABL* (the Philadelphia chromosome) biomarker to determine if patients with chronic myeloid leukemia should be treated with imatinib (Druker et al. 2001; Chen et al. 2016).

6.3.2 Barriers to Clinical Translation

Despite several pharmacogenetic discoveries that have been successfully implemented in clinical practice, therapeutic implementation of pharmacogenetic discoveries severely lags behind the current research (Manolio et al. 2013). There are several barriers that currently prevent the rapid and widespread implementation of pharmacogenetics.

One of the barriers is that there has been a lack of guidance on translating pharmacogenetic research discoveries to actionable decisions in clinical practice. The PGRN's Translational Pharmacogenetics program is a coordinated effort among multiple organizations to overcome the challenges preventing the implementation of pharmacogenetic-guided therapy (Shuldiner et al. 2013). The PGRN and PharmGKB have partnered to form the CPIC to provide clear, curated, and peer-reviewed pharmacogenetic guidelines to physicians to better inform drug prescribing decisions based on genotyping test results. The CPIC provides a list of gene–drug pairs along with guidelines for clinical practice that have been peer-reviewed and published (Relling and Klein 2011; Caudle et al. 2014; see https://cpicpgx.org/genes-drugs). The PharmGKB website lists dosing guidelines for drugs based on the patient's genotype, which consists of the CPIC published guidelines and tabulated genotype and phenotype descriptions along with therapy recommendations for each gene variant associated with the drug response (PharmGKB 2017b). PharmGKB (2017a) and the US FDA (2017) websites list FDA-approved drugs with pharmacogenetic information, which is included in the drug label.

Another barrier to implementation is the absence of infrastructure. A shared informational and policy infrastructure is needed for the implementation of genomic medicine. Patient genetic information must be accessible to healthcare providers easily and in a timely manner to be useful. Currently, genetic tests are performed at the time of prescribing a drug, which requires turnaround times of at least several days, which may not be practical in some cases. The infrastructure must be able to handle big data and must be secure to protect patient privacy. Currently, the healthcare system in the United States is not integrated into a common infrastructure and different health institutions use different patient management portals which do not integrate with each other. Thus, information-sharing is not possible when patients use multiple providers or change providers during their lifetimes. Several organizations are forming partnerships to address this problem. The National Institutes of Health's (NIH) PMI Working Group is creating a central repository for the volunteers involved in the PMI program. This central repository will collect, store, process, analyze, and retrieve the information of all biological specimens, including DNA, RNA, plasma, microbiome, EHR data, patient survey data, and data collected from sensors or mobile technologies for the PMI cohort throughout the length of the program (PMI Working Group 2015). The eMERGE-PGx project, a partnership of eMERGE and the PGRN, has implemented an EHR-based repository of specific pharmacogenetic genotypes, clinical phenotype data, and CDS modules for ~9000 patients which will be shared across 10 academic medical centers and health systems in the emERGE-II network to promote pharmacogenomics discoveries and their implementation into clinical practice (Rasmussen-Torvik et al. 2014). The Mayo Clinic Center of Individualized Medicine (CIM) has also implemented an integrated EMR-based infrastructure across multiple sites to aid the translation of pharmacogenomic discoveries to clinical practice (Farrugia and Weinshilboum 2013).

A major barrier to widespread implementation of personalized medicine is clinician resistance. This resistance stems from lack of education in clinicians, differing views about personalized medicine and pharmacogenetic results, and difficulty in interpretation of pharmacogenetic data into actionable clinical decisions (Petersen et al. 2014). The problems of interpretation and ambiguity are addressed by CDS modules and guidelines such as those provided by the CPIC. Physicians trained in pharmacogenomics are inclined to be early adopters of personalized medicine (Stanek et al. 2012). Using the early adopters to champion the cause of pharmacogenetics, conducting pilot projects, employing genetic counselors in clinical settings, and obtaining institutional commitment can accelerate the implementation of genomic medicine in the clinic (Manolio et al. 2013).

The cost and reimbursement of pharmacogenetic testing is also a barrier to implementation. Lack of evidence of a significant effect of pharmacogenetic tests to treatment efficacy is often cited as a reason for inadequate reimbursement (Cohen et al. 2013). As the benefits in efficacy and safety of drug therapy, as described in the previous section, become widely known among the healthcare community, it will become quite evident that the cost of the trial-and-error method of drug prescription is much higher than the cost of genotyping. Also, with the cost of genotyping decreasing dramatically, pharmacogenetic tests will become more economical. The various challenges to implementing pharmacogenomics in clinical practice and their solutions are described in detail by Manolio et al. (2013)

6.4 CLINICALLY RELEVANT PHARMACOGENOMICS MOVING FORWARD

While there are many barriers to the widespread use of pharmacogenomics in clinical practice, several efforts are being made to overcome these barriers by various research and government organizations. The PGRN has established the Translational Pharmacogenetics Program to accelerate the implementation of pharmacogenetics in clinical practice (Shuldiner et al. 2013). The NIH's Precision Medicine Initiative has been launched involving multiple academic, industry, and government organizations to improve patient care in the United States (PMI Working Group 2015). Here, we discuss some of the next steps in pharmacogenomics for its successful implementation in clinical practice.

6.4.1 Implementation

The current clinical implementation of pharmacogenetics is a reactive approach involving a one-drug–one-gene model, where a genetic test is ordered at the time of prescribing a drug. This is inefficient, expensive, and causes a delay in treatment, which can be improved. The alternative is a proactive approach involving a preemptive one-time genotyping with test results stored in an EHR integrated with CDS available to the physician at the time of treatment to deliver genomically guided therapy. This approach overcomes several of the disadvantages of the reactive pharmacogenetics testing, such as delay in obtaining test results and cost of testing. With the rapidly decreasing cost of genotyping, one-time genotyping of all pharmacogenetic variants is also more cost efficient than multiple one-gene tests ordered before each drug prescription. Several studies have already shown that a preemptive genotyping program integrated with EMRs is more efficient than the reactive model and could also improve patient safety (Schildcrout et al. 2012; Bielinski et al. 2014). Several research and governmental organizations have contributed to providing the infrastructure required for this preemptive pharmacogenomics model, such as the FDA's list of pharmacogene–drug pairs (US FDA 2017), guidelines and actionable clinical decisions for drug prescribing based on pharmacogenetic test results (Relling and Klein 2011; Whirl-Carrillo et al. 2012), prototype implementations as proof of concepts (Rasmussen-Torvik et al. 2014), and implementation and best practices guidelines (Shuldiner et al. 2013). The PGRN's Translational Pharmacogenetic Program recruited multiple sites to implement a preemptive pharmacogenetics model (Shuldiner et al. 2013). The progress of the programs in five US medical centers – the Mayo Clinic, Mount Sinai Medical Center, St. Jude Children's Research Hospital, University of Florida, and Vanderbilt University Medical Center – is reviewed by Dunnenberger et al. (2015). These early-adopter sites will provide the critical guidance on the process for clinical implementation of a proactive pharmacogenetics program to other sites throughout the country.

The regulatory process is a challenge for clinical implementation. Typically, translation of any biomarker can require stringent evidence from RCTs, which may not always be possible due to cost or infeasibility, as in the case of rare adverse events. The strict process is also true for labeling of drugs with pharmacogenomic biomarkers. Only approximately 13–15 percent of the drugs approved by the US FDA contain pharmacogenomic information in their labeling. The CPIC's list of actionable gene–drug pairs is also relatively small compared to the known actionable pharmacogenetic discoveries (Relling and Evans 2015). It has been proposed that other types of evidence should be considered for translation, such as observational studies at hospitals, trial programs in pharmacogenetics or data mining from biobanks linked to EMRs, but this is an active area of discussion in the field (Crawford et al. 2014).

The education and training of clinicians in the utilization of pharmacogenetic discoveries is crucial to the adoption and success of pharmacogenomics in healthcare. A collaborative effort from researchers, clinicians, genetic counselors, and government and healthcare officials is required to make personalized medicine a success (Manolio et al. 2013). While the cost of genotyping is decreasing rapidly, reimbursement for these costs varies across providers. Demonstrating that the savings due to genomically guided medicine are substantial is required to convince the patients and the healthcare community of the value of personalized medicine (Manolio et al. 2013). For example, patients carrying the loss-of-function allele in the *CYP2C19* gene are poor metabolizers of the drug clopidogrel and hence must be prescribed alternative therapy such as ticagrelor, which costs ~$250 for a 30-day supply (Harper and Topol 2012). However, patients who carry the functional alleles in the *CYP2C19* gene are ultra-rapid or extensive metabolizers and can be prescribed clopidogrel, which costs ~$15–35 for a 30-day supply (Harper and Topol 2012). If the healthcare system invests in paying for preemptive pharmacogenetic testing leading to correct therapy selection, this can result in significant savings for the provider and the patient. Similarly, the use of pharmacogenetic testing for correct dose selection and prevention of adverse events can result in fewer office or hospital visits and significant cost benefits (Harper and Topol 2012).

6.4.2 Applications and Active Areas of Research

Applications of pharmacogenomics extend beyond patient treatment in a clinical setting. Here we review some of these applications that could have promising results in the near future. We also discuss active areas of research in the field.

6.4.2.1 Drug Development

Pharmacogenetic testing can be just as useful in drug development as it is in patient treatment. Instead of enrolling phenotypically similar individuals in an RCT, enrollment can be based on genomic screening of individuals, where only individuals carrying the candidate genomic markers involved in the mechanism of action of the drug being tested are enrolled. Several targeted anticancer drugs have been codeveloped with companion diagnostic testing. For example, trastuzumab, a monoclonal antibody, was developed to target the *HER2* gene, which is overexpressed in tumor cells in certain breast cancers (Vogel et al. 2002; Vogel and Franco 2003). Another example is the drug gefitinib, a tyrosine kinase inhibitor, which targets specific mutations in the *EGFR* gene in tumor cells in certain nonsmall-cell lung cancers (NSCLC) (Lynch et al. 2004). As compared to RCTs, genomically guided clinical trials can be more efficient and cost-effective, provide sufficient statistical power with smaller sample sizes, and offer better evidence for the regulatory process (McCarthy et al. 2005; Harper and Topol 2012).

6.4.2.2   Epistatic Interactions

The current implementation of pharmacogenetics-based therapy typically involves testing for single genes or biomarkers, as seen in the US FDA's or the CPIC's gene–drug list. However, pharmacogenomic traits are complex traits influenced by multiple genes or SNPs, their interactions with each other, and their interactions with the environment. The trait-associated SNPs identified in the large number of GWAS only explain a small proportion of variation in various traits. This is one component of the "missing heritability" problem (Sadee 2012). One of the causes of this unexplained variation could be gene–gene or gene–environment interaction (Sadee 2012). One of the environmental factors in pharmacogenomic studies is the administered drug, but several other factors, such as diet, comorbidities, smoking or drinking status, etc., could also influence the drug response phenotype. Epistatic interactions are particularly important in cancer treatment, where the complex and constantly evolving nature of tumor genomes often leads to drug resistance. For example, resistance to gefitinib, a tyrosine kinase inhibitor targeting the *EGFR* gene, in lung cancer patients has been found to be due to an additional *EGFR* mutation that is acquired in the presence of the drug (Pao et al. 2005). Pharmacogenomic analyses of tumor samples from cancer patients showing the drug-resistance phenotype or disease progression even with anticancer therapy could lead to the identification of the causal epistatic interactions and the development of novel targeted therapies (Weigelt and Reis-Filho 2014). Statistical analyses methods for the detection of such epistatic interactions have been reviewed earlier in this chapter (Motsinger et al. 2007).

6.4.2.3 Drug Synergy

For several complex diseases such as cancer, AIDS, and cardiovascular disorders, patients are prescribed two or more drugs. The main goals of such multiple drug regimens are to increase efficacy, reduce toxicity, and minimize drug resistance. However, drug–drug interactions can alter the effectiveness of the individual drugs. For example, ingesting multiple drugs that are metabolized by the hepatic cytochrome 2D6 (CYP2D6) enzyme changes the efficacy of the individual drugs (Monte et al. 2014). Predicting the effect of combinatorial drug therapies (i.e. drug synergy) is complex and arduous due to the large combinatorial space of drug combinations, dosage concentrations, and gene targets. Nevertheless, several dose–response methods have been developed to model synergy, such as the Loewe additivity method (Loewe 1953), the Bliss independence method (Bliss 1939), and the Chou–Talalay method (Chou 2010). These and other models are implemented in software programs such as Combenefit, which allow for the analysis and visualization of the effect of combination drug therapy using dose–response data (Di Veroli et al. 2016). Systems biology modeling, such as pathway or network analysis (discussed earlier in this chapter) can be used to identify the mechanisms involved in the gene–drug interaction and the drug–drug epistatic interactions (Yeh et al. 2009; Chen et al. 2015). Several open community-based or crowd-sourcing-based challenges have been launched to build computational models for synergy prediction using monotherapy and combination therapy dose response data along with molecular and genomic data (Bansal et al. 2014; Synapse 2015). The use of pharmacogenomics to discover drug synergy could lead to the development of targeted combination drug therapies that optimize efficacy and minimize toxicity for the treatment of drug-resistance diseases such as cancer and AIDS.

6.4.3 Ethical, Social, and Economic Issues

As described earlier in this chapter, EMRs are used to store patient genotypes, demographics, medical history, physiological conditions, and health outcomes. They are necessary in the age of big data and provide a rich source of data for pharmacogenomic analyses that can be computationally analyzed and shared across multiple sites. The use of EMRs raises the issue of control and privacy of this data and for patients. Patient data in EMRs or other sources must be de-identified before analyses. Protocols need to be established to maintain the security and privacy of the data and to prevent unintended use of the data (Kahn 2011; Kulynych and Greely 2017).

Ethical and social concerns also arise with the availability and accessibility of genomic data. There has been concern that health insurance companies could use genetic test results to identify "high-risk" individuals – that is, if they carry genetic variations that make them susceptible to certain diseases or conditions – and either decline them coverage or increase the cost of their premiums. There has also been concern about employers or corporations, if privy to an individual's genetic information, using this information to make hiring decisions (Guttmacher et al. 2003). To alleviate these concerns, the US Equal Employment Opportunity Commission (EEOC) has passed the Genetic Information Nondiscrimination Act of 2008 (US EEOC 2017), which prohibits the discrimination of individuals by employers or by health insurance providers on the basis of genetic information. Medical researchers and clinicians also face the ethical dilemma regarding whether they should inform patients about secondary findings from genetic testing. Efforts have been made to provide recommendations, but no specific protocol has yet been established (Green et al. 2013; Kocarnik and Fullerton 2014).

Economic challenges such as costs and reimbursement of genetic testing are also an issue. Currently, pretreatment genetic testing is either not reimbursed or has limited reimbursement under many healthcare systems (Cohen et al. 2013; Relling and Evans 2015).

Regulatory processes need to be instituted to ensure healthcare coverage for actionable pharmacogenetic tests to facilitate their acceptance and use in clinical practice (Manolio et al. 2013; Carr et al. 2014; Levy et al. 2014).

6.4.4 Resources

Various academic, government, and medical institutions are working toward the aim of personalized medicine individually and collaboratively by sharing data and resources, promoting and facilitating pharmacogenomics research, providing infrastructure and expertise, establishing required standards, and providing guidelines for implementation of pharmacogenomics in clinical practice. In Table 6.1 we have assembled a list of several institutions, consortia, and online resources available for pharmacogenomics research.

6.5 CONCLUSION

6.5.1 Future Directions

We are now in the era of big data and multi-omics data, which consist of extremely large volumes of data from multiple "-omic" sources such as the epigenome, the transcriptome, the proteome, and the metabolome, in addition to the genome. Information from these multi-omic sources can be integrated to paint a more complete picture of the variability in individual response to a drug. These fields have been used to find genetic factors involved in various complex diseases and traits, but have only recently been applied to the drug response phenotype and have yet to be widely used in clinical practice. While the use of genomic data has identified, and continues to identify, several genetic factors influencing the variability in drug response, it cannot explain all the variation in drug response phenotypes. This is because the drug response is a complex trait that is influenced by several factors such as age, drugs, comorbidities, diet, and environment, in addition to the genome. The genome is a static entity, however, the other "-omes" are dynamic in nature and are

Table 6.1 *A list of institutions, consortia, and online resources related to pharmacogenomics research*

	Resource	URL
Institutions	National Institutes of Health (NIH)	www.nih.gov
	National Human Genome Research Institute (NHGRI)	www.genome.gov
	European Bioinformatics Institute (EBI)	www.ebi.ac.uk
	National Cancer Institute (NCI)	www.cancer.gov
	National Center for Biotechnology Information (NCBI)	www.ncbi.nlm.nih.gov
Consortia	Pharmacogenomics Research Network (PGRN)	www.pgrn.org
	Pharmacogenomics Research Network – RIKEN (PGRN-RIKEN)	www.pgrn.org/pgrn-riken.html
	Pharmacometabolomics Research Network (PMRN)	https://pharmacometabolomics.duhs.duke.edu
	Clinical Pharmacogenetics Implementation Consortium (CPIC)	https://cpicpgx.org
	Genomic Medicine Alliance (GMA)	www.genomicmedicinealliance.org
	Ubiquitous Pharmacogenomics (U-PGx)	http://upgx.eu
	International Warfarin Pharmacogenetics Consortium (IWPC)	www.pharmgkb.org/page/iwpc

Table 6.1 (*cont.*)

Resource	URL
International Warfarin Pharmacogenetics Consortium – Genome Wide Association Studies (IWPC-GWAS)	www.pharmgkb.org/page/ iwpc-gwas
International Clopidogrel Pharmacogenomics Consortium (ICPC)	www.pharmgkb.org/page/ icpc
International SSRI Pharmacogenomics Consortium (ISPC)	www.pharmgkb.org/page/ ispc
International Tamoxifen Pharmacogenomics Consortium	www.pharmgkb.org/page/ itpc
Consortium on Breast Cancer Pharmacogenomics (COBRA)	http://medicine.iupui.edu/ clinpharm/COBRA/
Drug-Induced Liver Injury Network (DILIN)	www.dilin.org
Genome-based Therapeutic Drugs for Depression (GENDEP)	http://gendep.iop.kcl.ac.uk/ results.php
International Consortium on Lithium Genetics (ConLiGen)	www.conligen.org
Precision medicine initiatives	All of Us Research program (NIH's Precision Medicine Initiative)	www.nih.gov/research-training/allofus-research-program
	Center for Individualized Medicine (CIM) at the Mayo Clinic	http://mayoresearch.mayo .edu/center-for-individualized-medicine/
	eMERGE-PGx project	https://emerge.mc .vanderbilt.edu/projects/ emerge-pgx/

Table 6.1 (*cont.*)

	Resource	URL
Gene–drug pairs information	Pharmacogenomics Knowledgebase (PharmGKB) drug labels	www.pharmgkb.org/view/drug-labels.do
	US FDA Table of Pharmacogenomic Biomarkers in Drug Labeling	www.fda.gov/Drugs/ScienceResearch/ResearchAreas/Pharmacogenetics/ucm083378.htm
	DrugBank	www.drugbank.ca
Genomic databases	Genome-wide association studies (GWAS) catalog	www.ebi.ac.uk/gwas/
	Phenome-wide association studies (PheWAS) catalog	https://phewascatalog.org
	Single Nucleotide Polymorphism database (dbSNP)	www.ncbi.nlm.nih.gov/projects/SNP/
	Database of Genotypes and Phenotypes (dbGaP)	www.ncbi.nlm.nih.gov/gap
	1000 Genomes project	www.internationalgenome.org
Pathway databases	Kyoto Encyclopedia of Genes and Genomes (KEGG)	www.kegg.jp
	Molecular Signatures Database (MSigdb)	http://software.broadinstitute.org/gsea/msigdb
	Network Data Exchange (NDEx)	www.ndexbio.org/#/
	Reactome	www.reactome.org
Electronic health records	Electronic Medical Records and Genomics (eMERGE)	https://emerge.mc.vanderbilt.edu
	BioVU	https://victr.vanderbilt.edu/pub/biovu/

Table 6.1 (*cont.*)

Resource	URL
BioBank Japan	https://biobankjp.org/ english/leaflet/index.html
Personalized Medicine Research Project (PMRP)	www.marshfieldresearch .org/chg/pmrp

affected by several of the aforementioned environmental factors. Thus, the integration of data from multiple "-omes" could aid in elucidating the unexplained variation or "missing heritability" for the dose–response phenotypes.

Transcriptomics is the analysis of all the RNA transcripts in a population of cells. RNA transcripts can be quantified using methods such as RNA sequencing (RNA-Seq), reverse transcription polymerase chain reaction (RT-PCR) or quantitative real-time polymerase chain reaction (qPCR). RNA expression data can be measured once or both before and after drug exposure to examine changes in RNA expression associated with drug response. Several noncoding RNAs (ncRNAs) have been associated with complex diseases such acute myeloid leukemia (AML) and acute promyelocytic leukemia (Valleron et al. 2012) and hepatocellular carcinoma (Luk et al. 2011; Xu et al. 2014). Similarly for the dose–response phenotype, studies have found that specific microRNAs (miRNAs) are associated with drug-induced liver injury (Starkey Lewis et al. 2011) and drug-induced TEN (Ichihara et al. 2014). Gene expression profiling can also be used to discover the mode of action or target of a drug as part of drug development. Lopez et al. identified miR-1202 as a potential target for novel anti-depressant drugs based on its differential expression in individuals with depression (Lopez et al. 2014). The results from these studies indicate the use of ncRNAs as potential biomarkers for drug response phenotypes.

Proteomics refers to the study of the entire set of proteins and their abundance levels in an organism. Proteins regulate various cellular processes and are involved in the dynamic and kinetic responses to drug treatments, positioning them closer to the drug response phenotype as compared to DNA or RNA. Thus, the integration of proteomics and genomics can give crucial insight into the mechanism of action of cellular responses to drug treatments, leading to the field of "pharmacoproteomics" (Jain 2004). Various immunoassays such as Co-Immunoprecipitation (Co-IP) assay or Affinity-based purification assays can be used for a targeted approach to capture proteins of interest, or all proteins in a cell can be used for analyses using techniques such as mass spectrometry. The pharmacoproteomics approach was successfully used to identify the synergistic effect of combination therapy consisting of PU-H71 and ibrutinib in activated B-cell-like diffuse large B-cell lymphoma (ABC DLCBL) (Goldstein et al. 2015).

Epigenomics is the study of the epigenetic modifications to the DNA in a cell. Epigenetic marks vary across different cell types, are modified during cellular development, and can also change due to aging, thus altering the gene expression in those cells. Epigenetic alterations in genes involved in DNA damage response pathways and cell cycle control pathways have been implicated in tumorigenesis. These epigenetic alterations can be used as biomarkers of drug response. Several drugs targeting the epigenome have been approved by the FDA – for example, the drugs azacitidine and decitabine, which inhibit the activity of DNA methyl transferases are used in the treatment of myelodysplastic syndrome (MDS) or AML (Fenaux et al. 2009; Lübbert et al. 2016). Since cancer encompasses both the genome and the epigenome, incorporating epigenomic analyses into cancer drug discovery and development is crucial (Jones et al. 2016).

Metabolomics involves the study of small molecules, metabolites, to determine the metabolic state of an individual. The metabolic signature of an individual reflects both the genetic and environmental influences on the phenotype and can provide a wealth of information that can be used to guide pharmacogenomics analyses. Pharmacometabolomics can inform pharmacogenomic research to identify biomarkers associated with drug response phenotypes based on individual metabolic signatures (Neavin et al. 2016). Changes in metabolic profiles after drug exposure can be used to predict variation in drug efficacy. The metabotype or the metabolic signature of a patient can be used to classify or stratify patients during pharmacogenomic analyses or drug development in a clinical trial. Since the metabolic state represents thousands of metabolites and their interactions with each other and the drug, it provides a closer and more dynamic view of the biochemical processes within a cell or tissue and can thus be used to identify the underlying molecular mechanisms of dose–response phenotypes (Kaddurah-Daouk and Weinshilboum 2014; Beger et al. 2016). An integrated approach using both genomics and metabolomics was used to identify biomarkers associated with variability in hydrochlorothiazide response in patients diagnosed with hypertension (Shahin et al. 2016).

The aforementioned "-omics" are complementary to pharmacogenomics, and the use of these large datasets requires a systems analysis approach. Various systems approaches, including network and pathway analysis approaches, can be used to identify and summarize the complex biological relationships and discover the mechanisms underlying the variability in drug response (Wang 2010; Khatri et al. 2012; Krumsiek et al. 2012; Kaddurah-Daouk et al. 2015). The integration of other "-omics" to pharmacogenomics has the potential to identify the yet unknown genetic and environmental factors influencing variability in drug response, and thus revolutionize precision medicine.

6.5.2 Summary

Pharmacogenomics is an exciting area of gene mapping – with the overall goal of connecting genetic variation with variation in drug response. Detecting and understanding such associations has great promise for improving patient care. We have reviewed some of the main resources, tools, study designs, and statistical methods used in pharmacogenomics research.

We are presented with a wealth of data, knowledge, resources, and expertise for pharmacogenomics research. We have also highlighted advantages and disadvantages in pharmacogenomics in contrast to other applications in complex trait mapping. We have highlighted a number of translational successes, and discussed some of the challenges in advancing additional findings. As the field progresses, advances in biotechnology, computer science, statistical methods, and bioinformatics, along with the efforts of research and government organizations and the cooperation, education, and awareness of clinicians and patients, will lead us into the era of personalized medicine.

REFERENCES

1000 Genomes Project Consortium, 2015. A global reference for human genetic variation. *Nature*, 526(7571):68–74.

Akaike H, 1974. A new look at the statistical model identification. *IEEE Trans Autom Control*, 19:716–723.

Ashburner M, Ball CA, Blake CA, et al., 2000. Gene ontology: tool for the unification of biology. *Nat Genet*, 25(1):25–29.

Bachtiar M, Lee CGL, 2013. Genetics of population differences in drug response. *Curr Genet Med Rep*, 1(3):162–170.

Bansal M, Yang J, Karan C, et al., 2014. A community computational challenge to predict the activity of pairs of compounds. *Nat Biotechnol*, 32(12):1213–1222.

Beger RD, Dunn W, Schmidt M, et al., 2016. Metabolomics enables precision medicine: a white paper, community perspective. *Metabolomics*, 12(9):149.

Benjamini Y, Hochberg Y, 1995. Controlling the false discovery rate: a practical and powerful approach to multiple testing. *J Roy Soc B*, 57(1):289–300.

Bielinski SJ, Olson M, Pathak JE, et al., 2014. Preemptive genotyping for personalized medicine: design of the right. *Mayo Clin Proc*, 89(1):.

Brown CC, Havener TM, Medina MW, et al., 2012a. A genome-wide association analysis of temozolomide response using lymphoblastoid cell lines shows a clinically relevant association with MGMT. *Pharmacogenet Genomics*, 22 (11):796–802.

Brown CC, Havener TM, Medina MW, et al., 2012b. Multivariate methods and software for association mapping in dose–response genome-wide association studies. *BioData Min*, 5(1):21.

Brown CC, Havener TM, Medina MW, et al., 2014. Genome-wide association and pharmacological profiling of 29 anticancer agents using lymphoblastoid cell lines. *Pharmagenomics*, 15(2):137–146.

Burdett T, et al., 2017 The NHGRI-EBI catalog of published genome-wide association studies. www.ebi.ac.uk/gwas.

Burmester JK, Sedova M, Shapero MH, Mansfield E, 2010. DMETTM microarray technology for pharmacogenomics-based personalized medicine. *Methods Mol Biol (Clifton NJ)*, 632:99–124.

Bliss CI, 1939. The toxicity of poisons applied jointly. *Ann Appl Biol*, 26(3):585–615.

Cantor RM, Lange K, Sinsheimer JS, 2010. Prioritizing GWAS results: a review of statistical methods and recommendations for their application. *Am J Hum Genet*, 86(1):6–22.

Caraco Y, Blotnick S, Muszkat M, 2008. CYP2C9 genotype-guided warfarin prescribing enhances the efficacy and safety of anticoagulation: a prospective randomized controlled study. *Clin Pharmacol Ther*, 83(3):460–470.

Carr DF, Alfirevic A, Pirmohamed M, 2014. Pharmacogenomics: current state-of-the-art. *Genes (Basel)*, 5(2):430–443.

Caudle KE, Klein TE, Hoffman JM, et al., 2014. Incorporation of pharmacogenomics into routine clinical practice: the Clinical Pharmacogenetics Implementation Consortium (CPIC) guideline development process. *Curr Drug Metab*, 15 (2):209–217.

Che R, Jack JR, Motsinger-Reif AA, Brown CC, 2014. An adaptive permutation approach for genome-wide association study: evaluation and recommendations for use. *BioData Min*, 7:9.

Chen D, Liu X, Yang Y, Yang H, Lu P, 2015. Systematic synergy modeling: understanding drug synergy from a systems biology perspective. *BMC Syst Biol*, 9.

Chen P, Lin JL, Lu CS, et al., 2011. Carbamazepine-induced toxic effects and HLA-B*1502 screening in Taiwan. *N Engl J Med*, 364(12):1126–1133.

Chen S, Sutiman N, Chowbay B, 2016. Pharmacogenetics of drug transporters in modulating imatinib disposition and treatment outcomes in chronic myeloid leukemia and gastrointestinal stromal tumor patients. *Pharmacogenomics*, 17 (17):1941–1955.

Chou TC, 2010. Drug combination studies and their synergy quantification using the Chou-Talalay method. *Cancer Res*, 70(2):440–446.

Choy E, Yelensky R, Bonakdar S, et al., 2008. Genetic analysis of human traits in vitro: drug response and gene expression in lymphoblastoid cell lines. *PLoS Genet*, 4(11):e1000287.

Cohen J, Wilson A, Manzolillo K, 2013. Clinical and economic challenges facing pharmacogenomics. *Pharmacogenomics J*, 1363(10):378–388.

Cox NJ, Gamazon ER, Wheeler HE, Dolan ME, 2012. Clinical translation of cell-based pharmacogenomic discovery. *Clin Pharmacol Ther*, 92(4):425–427.

Crawford DC, Crosslin DR, Tromp G, et al., 2014. EMERGEing progress in genomics-the first seven years. *Front Genet*, 5:1–11.

Dahlin A, Denny J, Roden DM, et al., 2015. CMTR1 is associated with increased asthma exacerbations in patients taking inhaled corticosteroids. *Immunity Inflamm Dis*, 3(4):350–359.

Dausset J, Cann H, Cohen D, et al., 1990. Centre d'etude du polymorphisme humain (CEPH): collaborative genetic mapping of the human genome. *Genomics*, 6(3):575–577.

Delaney JT, Ramirez EB, Pulley JM, et al., 2012. Predicting clopidogrel response using DNA samples linked to an electronic health record. *Clin Pharmacol Ther*, 91(2):257–263.

Denny JC, Ritchie MD, Basford, MA, et al., 2010. PheWAS: demonstrating the feasibility of a phenome-wide scan to discover gene-disease associations. *Bioinformatics*, 26(9):1205–1210.

Denny JC, Crawford DC, Ritchie MD, et al., 2011. Variants near FOXE1 are associated with hypothyroidism and other thyroid conditions: using electronic medical records for genome- and phenome-wide studies. *Am J Hum Genet*, 89(4):529–542.

Denny JC, Bastarache L, Ritychie M, et al., 2013. Systematic comparison of phenome-wide association study of electronic medical record data and genome-wide association study data. *Nat Biotechnol*, 31(12):1102–1111.

Di Veroli GY, Fornari C, Wang D, et al., 2016. Combenefit: an interactive platform for the analysis and visualization of drug combinations. *Bioinformatics*, 32 (18):2866–2868.

Druker BJ, Talpaz M, Resta DJ, et al., 2001. Efficacy and safety of a specific inhibitor of the BCR-ABL tyrosine kinase in chronic myeloid leukemia. *N Engl J Med*, 344 (14):1031–1037.

Dunnenberger HM, Crews KR, Hoffman JM, et al., 2015. Preemptive clinical pharmacogenetics implementation: current programs in five US medical centers. *Annu Rev Pharmacol Toxicol*, 55:89–106.

Epstein RS, Moyer TP, Aubert RE, et al., 2010. Warfarin genotyping reduces hospitalization rates results from the MM-WES (Medco-Mayo Warfarin Effectiveness Study). *JAC*, 55:2804–2812.

Farrugia G, Weinshilboum RM, 2013. Challenges in implementing genomic medicine: the Mayo Clinic Center for Individualized Medicine. *Clin Pharmacol Ther*, 94(2):204–206.

Fenaux P, Mufti GJ, Hellstrom-Lindberg E, et al., 2009. Efficacy of azacitidine compared with that of conventional care regimens in the treatment of higher-risk myelodysplastic syndromes: a randomised, open-label, phase III study. *Lancet Oncol*, 10:223–232.

Giacomini KM, Yee SW, Mushiroda T, et al., 2016. Genome-wide association studies of drug response and toxicity: an opportunity for genome medicine. *Nat Rev Drug Discov*, 16.

Goldstein RL, Yang SN, Taldone T, et al., 2015. Pharmacoproteomics identifies combinatorial therapy targets for diffuse large B cell lymphoma. *J Clin Invest*, 125(12):4559–4571.

Green RC, Berg JS, Grody WW, et al., 2013. ACMG recommendations for reporting of incidental findings in clinical exome and genome sequencing. *Genet Med*, 15 (7):565–574.

Guttmacher AE, Collins FS, Clayton EW, 2003. Ethical, legal, and social implications of genomic medicine. *N Engl J Med*, 349(6):562–569.

Han B, Kang HM, Eskin E, 2009. Rapid and accurate multiple testing correction and power estimation for millions of correlated markers. *PLoS Genet*, 5(4). https://doi.org/10.1371/journal.pgen.1000456.

Harper AR, Topol EJ, 2012. Pharmacogenomics in clinical practice and drug development. *Nat Biotechnol*, 30(11):1117–1124.

Hetherington S, Hughes AR, Mosteller M, et al., 2002. Genetic variations in HLA-B region and hypersensitivity reactions to abacavir. *Lancet*, 359(9312):1121–1122.

Huang RS, Duan S, Bleibel W, et al., 2007. A genome-wide approach to identify genetic variants that contribute to etoposide-induced cytotoxicity. *Proc Natl Acad Sci*, 104(23):9758–9763.

Ichihara A, Wang Z, Jinnin M, et al., 2014. Upregulation of miR-18a-5p contributes to epidermal necrolysis in severe drug eruptions. *J Allergy Clin Immunol*, 133:1065–1074.

Illumina, 2017. VeraCode ADME Core Panel. www.illumina.com/content/dam/illumina-marketing/documents/products/datasheets/datasheet_veracode_adme_core_panel.pdf.

International HapMap Consortium, 2003. The International HapMap Project. *Nature*, 426(6968):789–796.

International Human Genome Sequencing Consortium, 2001. Initial sequencing and analysis of the human genome. *Nature*, 409(6822):860–921.

International Warfarin Pharmacogenetics Consortium, 2009. Estimation of the warfarin dose with clinical and pharmacogenetic data. *N Engl J Med*, 360(8):753–764.

Ioannidis JPA, Ntzani EE, Trikalinos TA, Contopoulos-Ioannidis DG, 2001. Replication validity of genetic association studies. *Nat Genet*, 29(3):306–309.

Jack J, Rotroff D, Motsinger-Reif AA, 2014. Cell lines models of drug response: successes and lessons from this pharmacogenomic model. *Curr Mol Med*, 14 (7):833–840.

Jain K, 2004. Role of pharmacoproteomics in the development of personalized medicine. *Pharmacogenomics*, 5(3):331–336.

Johnson JA, Caudle KE, Gong L, et al., 2017. Clinical Pharmacogenetics Implementation Consortium (CPIC) guideline for pharmacogenetics-guided warfarin dosing: 2017 update. *Clin Pharmacol Ther*, 102:397–404.

Johnson RC, Caudle KE, Gong L, et al., 2010. Accounting for multiple comparisons in a genome-wide association study (GWAS). *BMC Genomics*, 11(1):724.

Jones PA, Issa J-PJ, Baylin S, 2016. Targeting the cancer epigenome for therapy. *Nat Rev Genet*, 17(10):630–641.

Kaddurah-Daouk R, Weinshilboum RM, 2014. Pharmacometabolomics: implications for clinical pharmacology and systems pharmacology. *Clin Pharmacol Ther*, 95(2):154–167.

Kaddurah-Daouk R, Weinshilboum R, Pharmacometabolomics Research Network, 2015. Metabolomic signatures for drug response phenotypes: pharmacometabolomics enables precision medicine. *Clin Pharmacol Ther*, 98 (1):71–75.

Kahn SD, 2011. On the future of genomic data. *Science*, 331(6018):728–729.

Kanehisa M, Goto S, 2000. KEGG: Kyoto Encyclopedia of Genes and Genomes. *Nucleic Acids Res*, 28(1):27–30.

Khatri P, Sirota M, Butte AJ, Ten years of pathway analysis: current approaches and outstanding challenges. *PLoS Comput Biol*, 8(2):e1002375. https://doi.org/10 .1371/journal.pcbi.1002375.

Kocarnik JM, Fullerton SM, 2014. Returning pleiotropic results from genetic testing to patients and research participants. *JAMA*, 311(8):795–796.

Kontopantelis E, Reeves D, 2010. Metaan: random-effects meta-analysis. *Stata J*, 10 (3):395–407.

Kraft P, Cox DG, 2008. Study designs for genome-wide association studies. *Adv Genet*, 60:465–504.

Krumsiek J, Suhre K, Evans AM, et al., 2012. Mining the unknown: a systems approach to metabolite identification combining genetic and metabolic information. *PLoS Genet*, 8(10):e1003005.

Kulynych J, Greely HT, 2017. Clinical genomics, big data, and electronic medical records: reconciling patient rights with research when privacy and science collide. *J Law Biosci*, 15:94–132.

Laurie CC, Doheny KF, Mirel DB, et al., 2010. Quality control and quality assurance in genotypic data for genome-wide association studies. *Genet Epidemiol*, 34(6):591–602.

Leckband SG, Kelsoe JR, Dunnenberger HM, et al., 2013. Clinical Pharmacogenetics Implementation Consortium guidelines for HLA-B genotype and carbamazepine dosing. *Clin Pharmacol Ther*, 94(3):324–328.

Levy KD, Decker BS, Carpenter JS, et al., 2014. Prerequisites to implementing a pharmacogenomics program in a large health-care system. *Clin Pharmacol Ther*, 96(3):307–309.

Li M-X, Yeung JMY, Cherny SS, Sham PC, 2012. Evaluating the effective numbers of independent tests and significant p-value thresholds in commercial genotyping arrays and public imputation reference datasets. *Hum Genet*, 131 (5):747–756.

Loewe S, 1953. The problem of synergism and antagonism of combined drugs. *Arzneimittelforschung*, 3(6):285–290.

Lopez JP, Lim R, Cruceanu C, et al., 2014. mir-1202 is a primate-specific and brain-enriched microRNA involved in major depression and antidepressant treatment. *Nat Med*, 20(7):764–768.

Low S-K, Chung S, Takahashi A, et al., 2013. Genome-wide association study of chemotherapeutic agent-induced severe neutropenia/leucopenia for patients in Biobank Japan. *Cancer Sci*, 104(8):1074–1082.

Lübbert M, Suciu S, Hagemeijer A, et al., 2016. Decitabine improves progression-free survival in older high-risk MDS patients with multiple autosomal monosomies: results of a subgroup analysis of the randomized phase III study 06011 of the EORTC Leukemia Cooperative Group and German MDS Study Group. *Ann Hematol*, 95(2):191–199.

Luk JM, Burchard J, Zhang C, et al., 2011. DLK1-DIO3 genomic imprinted microRNA cluster at 14q32.2 defines a stemlike subtype of hepatocellular carcinoma associated with poor survival. *J Biol Chem*, 286 (35):30706–30713.

Lynch TJ, Bell D, Sordella R, et al., 2004. Activating mutations in the epidermal growth factor receptor underlying responsiveness of non-small-cell lung cancer to gefitinib. *N Engl J Med*, 350(21):2129–2139.

Manolio TA, Chisholm RL, Ozenberger B, et al., 2013. Implementing genomic medicine in the clinic: the future is here. *Genet Med*, 15(4):258–267.

Martin MA, Hoffman JM, Freimuth RR, et al., 2014. Clinical Pharmacogenetics Implementation Consortium Guidelines for HLA-B genotype and abacavir dosing: 2014 update. *Clin Pharmacol Ther*, 95(5):499–500.

Marty M, Cognetti D, Maraninci D, et al., 2005. Randomized phase II trial of the efficacy and safety of trastuzumab combined with docetaxel in patients with human epidermal growth factor receptor 2-positive metastatic breast cancer administered as first-line treatment: the M77001 study group. *J Clin Oncol*, 23(19):4265–4274.

McCarthy AD, Kennedy JL, Middleton LT, 2005. Pharmacogenetics in drug development. *Philos Trans R Soc Lond B Biol Sci*, 360(1460):1579–1588.

McCarty CA, Chisholm RL, Chute CG, et al., 2011. The eMERGE Network: a consortium of biorepositories linked to electronic medical records data for conducting genomic studies. *BMC Med Genomics*, 4(1):13.

Mega JL, Close SL, Wiviott SD, et al., 2009. Cytochrome P-450 polymorphisms and response to clopidogrel. *N Engl J Med*, 360:354–362.

Merico D, Gfeller D, Bader GD, 2009. How to visually interpret biological data using networks. *Nat Biotechnol*, 27(10):921–924.

Meta-Analysis, 2017. Comprehensive Meta-Analysis Software (CMA). www.meta-analysis.com.

Monte AA, Heard KJ, Campbell J, et al., 2014. The effect of CYP2D6 drug–drug interactions on hydrocodone effectiveness. *Acad Emerg Med*, 21(8):879–885.

Moore CB, Verma A, Pendergrass S, et al., 2015. Phenome-wide association study relating pretreatment laboratory parameters with human genetic variants in AIDS clinical trials group protocols. *Open Forum Infect Dis*, 2(1). https://doi.org/10.1093/ofid/ofu113.

Moore JH, 2003. The ubiquitous nature of epistasis in determining susceptibility to common human diseases. *Hum Hered*, 56:73–82.

Motsinger AA, Ritchie MD, Reif DM, 2007. Novel methods for detecting epistasis in pharmacogenomics studies. *Pharmacogenomics*, 8(9):1229–1241.

Motsinger AA, Dudek SM, Hahn LW, Ritchie MD, 2006. *Comparison of Neural Network Optimization Approaches for Studies of Human Genetics*. Berlin: Springer.

Motsinger-Reif AA, Jorgenson E, Relling MV, et al., 2013. Genome-wide association studies in pharmacogenomics: successes and lessons. *Pharmacogenet Genomics*, 23(8):383–394.

Neavin D, Kaddurah-Daouk R, Weinshilboum R, 2016. Pharmacometabolomics informs pharmacogenomics. *Metabolomics*, 12(7):1–6.

Nebert DW, 1999. Pharmacogenetics and pharmacogenomics: why is this relevant to the clinical geneticist? *Clin Genet*, 56(4):247–258.

Neuraz A, Chouchana L, Malamut G, et al., 2013. Phenome-wide association studies on a quantitative trait: application to TPMT enzyme activity and thiopurine therapy in pharmacogenomics. *PLoS Comput Biol*, 9(12):e1003405.

Nicolae DL, Gamazon E, Zhang W, et al., 2010. Trait-associated SNPs are more likely to be eQTLs: annotation to enhance discovery from GWAS. *PLoS Genet*, 6(4):e1000888.

Noble WS, 2009. How does multiple testing correction work? *Nat Biotechnol*, 27 (12):1135–1137.

Pao W, Miller V, Zakowski M, et al., 2004. EGF receptor gene mutations are common in lung cancers from and "never smokers" and are associated with sensitivity of tumors to gefitinib and erlotinib. *Proc Natl Acad Sci USA*, 101(36):13306–13311.

Pao W, Miller VA, Politi KA, et al., 2005. Acquired resistance of lung adenocarcinomas to gefitinib or erlotinib is associated with a second mutation in the EGFR kinase domain. *PLoS Med*, 2(3):e73.

Patnala R, Clements J, Batra J, 2013. Candidate gene association studies: a comprehensive guide to useful in silico tools. *BMC Genet*, 14(1):39.

Peters EJ, Motsinger-Reif A, Havener TM, et al., 2011. Pharmacogenomic characterization of US FDA-approved cytotoxic drugs. *Pharmacogenomics*, 12 (10):1407–1415.

Petersen KE, Prows CA, Martin LJ, Maglo KN, 2014. Personalized medicine, availability, and group disparity: an inquiry into how physicians perceive and rate the elements and barriers of personalized medicine. *Public Health Genomics*, 17(4):209–220.

Pharmacogenomics at work, 1998. Pharmacogenomics at work. *Nat Biotechnol*, 16 (10):885.

Pharmacogenomics Knowledge Base (PharmGKB), 2017a. Pharmacogenomics Knowledge Base. Drug Labels. www.pharmgkb.org/view/drug-labels.do.

Pharmacogenomics Knowledge Base (PharmGKB) 2017b. Pharmacogenomics Knowledge Base. Dosing Guidelines. www.pharmgkb.org/view/dosing-guidelines.do.

Pharmacogenomics Research Network (PGRN), 2017. PGRN HUB. www.pgrn.org.

PMI Working Group, 2015. The Precision Medicine Initiative cohort program: building a research foundation for 21st century medicine. Precision Medicine Initiative Working Group Report to Advisory Committee to the Director of the NIH.

Pratt D, Chen J, Welker D, et al., 2015. NDEx, the network data exchange. *Cell Syst*, 1:302–305.

Purcell S, Neale B, Todd-Brown K, et al., 2007. PLINK: a tool set for whole-genome association and population-based linkage analyses. *Am J Hum Genet*, 81 (3):559–575.

R Development Core Team, 2016. *R: A Language and Environment for Statistical Computing*. Vienna: R Foundation for Statistical Computing.

Rasmussen-Torvik LJ, Stallings SC, Gordon AS, et al., 2014. Design and anticipated outcomes of the eMERGE-PGx project: a multicenter pilot for preemptive pharmacogenomics in electronic health record systems. *Clin Pharmacol Ther*, 96(4):482–489.

Relling MV, Klein TE, 2011. CPIC: Clinical Pharmacogenetics Implementation Consortium of the Pharmacogenomics Research Network. *Clin Pharmacol Ther*, 89(3):464–467.

Relling MV, Evans WE, 2015. Pharmacogenomics in the clinic. *Nature*, 526 (7573):343–350.

Ringnér M, Ringner M, 2008. What is principal component analysis? *Nat Biotechnol*, 26(3):303–304.

Ritchie MD, Motsinger AA , 2005. Multifactor dimensionality reduction for detecting gene–gene and gene–environment interactions in pharmacogenomics studies. *Pharmacogenomics*, 6(5):823–834.

Ritchie MD, 2012. The success of pharmacogenomics in moving genetic association studies from bench to bedside: study design and implementation of precision medicine in the post-GWAS era. *Human Genetics*, 131(10):1615–1626.

Ritchie MD, White BC, Parker JS, Hahn LW, Moore JH, 2003. Optimization of neural network architecture using genetic programming improves detection and modeling of gene–gene interactions in studies of human diseases. *BMC Bioinformatics*, 4(1). https://doi.org/10.1186/1471-2105-4-28.

Roden D, Pulley JM, Basford MA, et al., 2008. Development of a large-scale de-Identified DNA biobank to enable personalized medicine. *Clin Pharmacol Ther*, 84(3):362–369.

Sadee W, 2012. The relevance of "missing heritability" in pharmacogenomics. *Clin Pharmacol Ther*, 92(4):428–430.

SAS, 2017. SAS 9.4 software. www.sas.com/en_us/software/sas9.html#94-c-to-a.

Schildcrout JS, Denny JC, Bowton E, et al., 2012. Optimizing drug outcomes through pharmacogenetics: a case for preemptive genotyping. *Clin Pharmacol Ther*, 92(2):235–242.

Schwarz G, 1978. Estimating the dimension of a model. *Ann Stat*, 62:461–464.

Scott SA, Sangkuhl K, Stein CM, et al., 2013. Clinical Pharmacogenetics Implementation Consortium Guidelines for CYP2C19 genotype and clopidogrel therapy: 2013 update. *Clin Pharmacol Ther*, 94(3):317–323.

Shahin MH, Gong Y, McDonough CW, et al., 2016. A genetic response score for hydrochlorothiazide use. *Hypertension*, 68(3):621–629.

Shuldiner AR, Relling MV, Peterson JF, et al., 2013. The Pharmacogenomics Research Network Translational Pharmacogenetics Program: overcoming challenges of real-world implementation. *Clin Pharmacol Ther*, 94(2):207–210.

Stanek EJ, Sanders CL, Taber KAJ, et al., 2012. Adoption of pharmacogenomic testing by US physicians: results of a nationwide survey. *Clin Pharmacol Ther*, 91(3):450–458.

Starkey Lewis PJ, Dear J, Platt V, et al., 2011. Circulating microRNAs as potential markers of human drug-induced liver injury. *Hepatology*, 54(5):1767–1776.

Stolberg HO, Norman G, Trop I, 2004. Randomized controlled trials. *Am J Roentgenol*, 183(6):1539–1544.

Subramanian A, Tamayo P, Mootha VK, et al., 2005. Gene set enrichment analysis: a knowledge-based approach for interpreting genome-wide expression profiles. *Proc Natl Acad Sci USA*, 102(43):15545–15550.

Synapse, 2015. AstraZeneca-Sanger Drug Combination Prediction DREAM Challenge – syn4231880. www.synapse.org/#!Synapse:syn4231880/wiki/235645.

Takeuchi F, McGinnis R, Bourgeois S, et al., 2009. A genome-wide association study confirms VKORC1, CYP2C9, and CYP4F2 as principal genetic determinants of warfarin dose. *PLoS Genet*, 5(3):e1000433.

US EEOC, 2017. Genetic Information Nondiscrimination Act of 2008. www.eeoc .gov/laws/statutes/gina.cfm.

US FDA, 2017. Table of pharmacogenomic biomarkers in drug labeling. www.fda.gov/ Drugs/ScienceResearch/ResearchAreas/Pharmacogenetics/ucm083378.htm.

Valleron W, Laprevotte E, Gautier EF, et al., 2012. Specific small nucleolar RNA expression profiles in acute leukemia. *Leukemia*, 26(10). https://doi.org/10 .1038/leu.2012.111.

Vogel CL, Franco SX, 2003. Clinical experience with trastuzumab (herceptin). *Breast J*, 9(6): 452–462.

Vogel CL, Cobleigh MA, Tripathy D, et al., 2002. Efficacy and safety of trastuzumab as a single agent in first-line treatment of HER2-overexpressing metastatic breast cancer. *J Clin Oncol*, 20(3):719–726.

Vogel F, 1959. Moderne Probleme der Humangenetik. In *Ergebnisse der Inneren Medizin und Kinderheilkunde*, Berlin: Springer, pp. 52–125.

Wang L, 2010. Pharmacogenomics: a systems approach. *Wiley Interdisc Rev Syst Biol Med*, 2(1):3–22.

Watson VG, Motsinger-Reif A, Hardison NE, et al., 2011. Identification and replication of loci involved in camptothecin-induced cytotoxicity using CEPH pedigrees. *PLoS One*, 6(5). https://doi.org/10.1371/journal.pone.0017561.

Weigelt B, Reis-Filho JS, 2014. Epistatic interactions and drug response. *J Pathol*, 232(2):255–263.

Welsh M, Mangravite L, Medina MA, et al., 2009. Pharmacogenomic discovery using cell-based models. *Pharmacol Rev*, 61(4):413–429.

Wheeler HE, Dolan ME, 2012. Lymphoblastoid cell lines in pharmacogenomic discovery and clinical translation. *Pharmacogenomics*, 13(1):55–70.

Wheeler HE, Gamazon ER, Stark AL, et al., 2013. Genome-wide meta-analysis identifies variants associated with platinating agent susceptibility across populations. *Pharmacogenomics J*, 13(1):35–43.

Whirl-Carrillo M, McDonagh EM, Hebert JM, et al., 2012. Pharmacogenomics knowledge for personalized medicine. *Clin Pharmacol Ther*, 92(4):414–417.

Wilke RA, Berg RL, Peissig P, et al., 2007. Use of an electronic medical record for the identification of research subjects with diabetes mellitus. *Clin Med Res*, 5(1):1–7.

Willer CJ, Li Y, Abecasis GR, 2010. METAL: fast and efficient meta-analysis of genomewide association scans. *Bioinformatics*, 26(17):2190–2191.

Xu G, Yang F, Ding CL, et al., 2014. Small nucleolar RNA 113-1 suppresses tumorigenesis in hepatocellular carcinoma. *Mol Cancer*, 13(1):216.

Yee SW, Momozawa Y, Kamatani Y, et al., 2016. Genomewide association studies in pharmacogenomics: meeting report of the NIH Pharmacogenomics Research Network-RIKEN (PGRN-RIKEN) collaboration. *Clin Pharmacol Ther*, 100 (5):423–426.

Yeh PJ, Hegreness MJ, Aiden AP, Kishony R, 2009. Drug interactions and the evolution of antibiotic resistance. *Nat Rev Microbiol*, 7(6):460–466.

Yesupriya A, Yu W, Clyne M, Gwinn M, Khoury MJ, 2008. The continued need to synthesize the results of genetic associations across multiple studies. *Genet Med*, 10(8):633–635.

Ziliak D, O'Donnell P, Im HK, et al., 2011. Germline polymorphisms discovered via a cell-based, genome-wide approach predict platinum response in head and neck cancers. *Transl Res*, 157(5):265–272.

Zuvich RL, Armstrong LL, Bielinksi SJ, et al., 2011. Pitfalls of merging GWAS data: lessons learned in the eMERGE network and quality control procedures to maintain high data quality. *Genet Epidemiol*, 35(8):887–898.

7 The Ethical Status of an AI

James A. Foster and Donald C. Wunsch II

7.1 INTRODUCTION

Our objective here is to explore the ethical status of artificial intelligence (AI) and to reflect on the consequent normative implications for engineers. We frame the discussion in the context of major ethical theories, in order to be explicit about what might give an "artificial" entity ethical standing. We have attempted to avoid anthropocentric assumptions.

It is not our place here to define AI or to limit the discussion to AIs as embodied in some particular way. We cannot dive into the vast and contentious literature about consciousness, free will, self-awareness, rationality, and embodiment. It suffices for our purpose that we leave the concept of AI intuitive and unanalyzed, surveying only possible arguments for AIs having ethical status. We invite the reader to use whatever they imagine to be a "true AI" when they need a concrete example.

As we begin our considerations, we do not enter into current speculation about whether the existence of AIs will be a good or bad thing, nor about the ethics of creating or using AIs. In particular, we ignore legal and regulatory questions about how humans should interact with AIs. We are more interested in the ethical status of the AIs themselves than in the ethical issues consequent on their relations with other ethical entities such as humans. To the extent that humans enter our discussion, it is from the perspective of AIs interacting with them, not vice versa.

In what follows, we assume that "true AIs" exist, and refer to them as AIs. This is purely to make the discussion easier. This begs the question of whether such things ever will exist, or what they will be like. But it simplifies the exposition and keeps us focused on our reflections about ethical status.

7.2 THEORY-SPECIFIC CONSIDERATIONS

7.2.1 Nihilism

It is possible that AIs in fact have no ethical status. This would be the case if AIs lack some characteristic essential for ethical standing. Since we are not presuming that AIs have any particular characteristics, it is impossible to reject ethical nihilism *a priori* from this essentialist point of view.

For example, one might reject the very possibility of AI ethical status, arguing that AIs are not alive, are purely mechanical, or that they lack empathy or rationality. But it is at least debatable whether AIs can be alive. In any case, life is not a necessary precondition to ethical status, since some living things (such as bacteria) do not obviously have ethical status. It isn't clear what "purely mechanical" means, or that creatures that do have ethical status, such as humans, are not also "purely" mechanical – there is considerable debate about the nature and limitations of human autonomy and freedom of choice. Nor is it obvious that empathy or rationality are necessary and sufficient for ethical standing, as witnessed by millennia of debate among ethical theorists. Therefore we will generally pass over essentialist arguments for or against the existence of AI ethics as too contentious for this chapter.

But suppose an essentialist argument for human ethical status carries over to AI ethical status. For example, suppose the posited essential characteristic is taken to be the basis for human ethics, and AIs share that characteristic. Examples might include rationality or autonomy. We do consider this possibility below. But such arguments can lead to unpleasant conclusions. If the argument is that AIs have the essential characteristic, there may still be other, nonhuman, ethical bases for AIs which deserve exploration. If we posit that AIs have the requisite characteristics, but not to a sufficient degree to merit ethical status, then one must ask at what significance level would an AI lose its standing. The answer should apply equally to humans, and so this argument could imply that (at least some) humans also lack ethical standing, or that some people have fuller ethical standing than others. This is similar to Peter Singer's argument that nonhuman animals have a stronger claim to ethical treatment than humans whose rationality or potential is flawed (Singer 2009). And so any essentialist argument leading to conditional nihilism for AIs is irrelevant to our purpose – we are not talking about those AIs, but only about the ones that do have ethical status. That is, we will assume that AIs have ethical status and ask what the basis for that claim might be.

One might also argue that AI ethics may be derivative from the extent of their relationship to humans (and other entities with ethical standing), and so have no independent ethical standing. For example, some subset of AIs would presumably play important supporting roles in ethically charged cases, therefore inheriting ethical consideration. For example, AIs will have important roles in safety, security, survivability, and sustainability. However, the justification that this role has ethical import depends on the assumption that it has ethical import for humans. Either the same reason that justifies standing for humans justifies it for AIs, or standing is purely a matter of relationship with humans. Since AIs can, and probably will, exist independently of humans, this would imply that some AIs would have standing and others would not, which is unreasonable. Suppose that an AI has standing because of its relationship with humans. Now imagine that AI is sent in a rocket into space, never to interact with humans again. Does it lose its ethical status? That would be an odd consequence.

One might reply that AI ethical standing depends on the type and extent of relations with humans, so that AIs playing less important roles with humans may lack or have reduced ethical standing. This argument assumes that these relations matter precisely because they matter to humans. But, if AI ethics are a matter of degrees dependent on relations with humans, then by the parity principle we presented above, human ethical standing must also depend to some extent on their relations with each other. This is the same slippery slope we rejected above.

The hypothesis that AI ethics may be associated with relations between ethical agents such as humans and other AIs is worth exploring further. It would be useful to pursue this from the perspective of care ethics, though we don't do that here.

The conclusion seems to be that AI ethics do not derive from essential characteristics or from relationships that humans. So, if AIs have ethical standing, these are not the (only) reasons. And we reject ethical nihilism since our project is not to determine whether AIs have ethical standing, but to evaluate the possible justifications for this standing, on the assumption that they do.

7.2.2   Divine Command

The divine command theory of ethics posits that the ethical status of actions is determined, directly or indirectly, by fiat from a "higher" authority. However, any such theory must address Plato's counter-argument in the Euthyphro: Is an action good (he actually said "pious") because a god says so, or does the god say so because it is good? In the former case, the ethical status of the action is arbitrary. In the latter, it derives from something other than divine command. In either case, the divine command theory is not an adequate ethical explanation.

But for AIs this argument doesn't have the same force. The first AIs will be human creations, presumably. So, it is at least plausible that their ethical status will derive from human fiat. However, when a creator dictates an ethical standard for their creation, a non-arbitrary creator must have a reason for choosing that standard rather than another. That reason is the effective ethical basis, not the dictate itself. For example, one might argue that divine command theory for humans is explained by natural law, which may be built upon axiomatic truths and extended through reasoning. This is the argument made by Locke and by the American Founding Fathers. But in this case, it isn't a divine command that is the basis for ethics, but rather natural law or the axioms that it presumes.

But suppose an AI's initial creator fails to explicitly give their creations an ethical foundation – as is likely to happen if engineers ignore the issue. The first AIs may still derive their ethics by observing their creators, so that the "divine command" is in fact the behavior of the creators, transmitted unintentionally (think of the first *Star Trek* movie). In this case, the AI ethical basis may have a divine command foundation, and still have no further ethical standard, defeating the Euthyphro argument. Is this an example of divine command from the perspective of the AI and AI nihilism from the perspective of the creator?

It is at least plausible that this form of divine command can be the basis of AI ethics. This is a bleak prospect. Human behavior is often not exemplary, to say the least, regardless of what the foundation of human ethics might be. Humans often fail to model ethical behavior even for their own offspring. It is unlikely that humans will change their behavior significantly before AIs emerge, even if a new obligation to do so emerges from considerations of AI ethics.

On the other hand, unintentional divine command ethics for AIs may be inappropriate even if the creators are saints. There may be deeper justifications for ethical AIs that trump imitation. So, we proceed to other ethical theories.

7.2.3   Consequentialism

Consequentialist ethical theories argue that the ethical foundation of actions is determined by their consequences. The most common consequentialist theory is utilitarianism, in which what matters is the net change in well-being for some set of agents. The various approaches to utilitarian ethics hinge on how to determine net well-being and who are the relevant agents. Negative utilitarianism posits that the primary metric is preventing (or minimizing) suffering. Other theories give primacy to increasing the amount of some positive metric. But it is unclear exactly what such metrics should be. Theories often choose a criterion by replacing one vague term with another, such as "well-being" with "happiness." The leading conse-quentialists were Bentham, Mill, and, more recently, Peter Singer.

Debate over how to measure consequence for AIs may turn on whether the AIs necessarily have affective states, such as suffering or happiness. It is not clear that these are necessary consequences or preconditions for "artificial" intelligence. We defer a discussion of AI happiness and well-being in a psychological sense until that debate is resolved. For now, we make the conservative assumption that AIs will not have emotional or affective states, so that this type of utilitarian-ism is not relevant to our current discussion. However, the relation-ship between affective states and ethical foundations is a rich one with a huge philosophical literature and deserves further study.

In any case, the utilitarianism approach to AI ethics is likely to be a durable one. Is there another type of hedonistic calculus, to use Bentham's term (1789), that is relevant to AIs? Perhaps ethical AI behavior should maximize the amount of knowledge, or the speed of calculations, or energy efficiency. Perhaps ethical AIs maximize the number of peta-bytes of data per kilowatt. This is a coherent theory, even if alien to humans.

What each of these ideas have in common is that a utility function can be defined. That is, consequentialism reduces ethics to optimization, and this is exactly what the most successful forms of AI to date are highly capable of. Particularly in the context of multi-agent systems, each AI will typically maximize its own utility function. Whether optimization criteria can be mapped by some anthropomorphic argument to human affective states is irrelevant. The utility used by the AI is sufficient.

The time frame for optimization is also relevant. An AI might learn something similar to human altruism in a consequentialist system if it interacts with other agents and its utility function is long-term. This is actually possible if the AI is capable of reasoning about the likely response of other agents, factoring their response into its calculations, and optimizing over time. This can happen for evolu-tionary computing or reinforcement learning approaches to the iter-ated prisoner's dilemma.

The second front on which utilitarian theories often found is the question of whose well-being matters. Should it be just the AIs? Should it include humans? Should it include only the agent in question, or other current agents, or even future agents? It would seem axiomatic that agents with ethical status ought to be treated ethically, so that AIs have ethical requirements with respect to humans. It is worthwhile, however, to separate the basis for AI interactions with humans from the basis for AI behavior in general. Perhaps human obligations are deontological, stemming perhaps from filial piety, while utilitarian ethics justifies other AI ethics.

A third problem with calculus is how one measures well-being, even when the relevant metrics are defined. Is physical suffering more important than mental anguish? Is dissatisfaction with what one has more relevant than the absence of goods? Should future generations' well-being be discounted? With regard to AIs, should humans even be considered, and if so what weight should consequences to them have? Furthermore, optimization as a foundation for ethics is likely to be multi-objective, since the most difficult issues often arise from trade-offs among several if not many measures of well-being. This brings in an entire literature of multi-objective optimization.

In short, it seems likely that the foundations for AI ethics will be consequentialist, given the centrality of optimization to AI algorithms. However, it is important to consider carefully which consequences matter, and how much.

7.2.4 Deontology

Deontological theory, from the ancient Greek for "duty" or "necessity," maintains that the basis for ethical actions are obligations or duties, which flow from some innate quality in the agent or from the pain of contradiction. Rationality or autonomy are the most commonly cited conditions that imply constraint. The most prominent exponent of deontological ethics was Kant, though of course there have been many others.

One foundation of ethical standing is Kant's categorical imperative, which he formulated in three different ways in his *Foundations of Metaphysics of Morals* (Kant 1785). His first formulation was that the soundness of an ethical position should be judged by its potential universality. That is, if it is logically consistent for all agents to take the same position along with the focal agent. If the action can be supported for an individual but not universally, then it is a rationalization and should be rejected. For example, one should not lie, because if everyone lied at the same time, without knowing this was the case, then communication would be impossible.

Kant also formulated his categorical imperative in terms of autonomy, saying that one ought always to treat other agents as ends in themselves, and not merely as a means. To do otherwise contradicts rational autonomy, which Kant argued is inherent in what it means to be an agent subject to ethics.

At a minimum, if one justifies AI ethics with Kant's deontology, then the principle of parity would require that AIs treat humans as beings with ethical standing, and vice versa. With the universality test, the AI should only act in such a way that the action can be generalized to both AIs and humans. This is problematic since some AI actions are likely to be impossible for humans, and what is obligatory should at least be possible.

This argues against the parity principle, or Kant's universality formulation of the categorical imperative. It seems clear that if AIs are ethical entities, then human ethics will need to take them into consideration. So, ignoring parity in deontology would seem to leave humans with obligations that AIs would not need to reciprocate. This is ominously similar to arguments that some humans are so different from others that they transcend their ethical obligations to others. It might be possible to relax Kant's universality principle, perhaps so that AIs must be able to universalize their actions only with respect to all other AIs. Again, this would have frightening consequences for relations between AIs and humans, since it would imply that they would be justified in ignoring humans.

Autonomy, on the other hand, seems a more reasonable ethical foundation since it could apply to both AIs and humans. The difficulty, of course, is defining autonomy. Are AIs autonomous in the same way as humans? Does AI autonomy mean the ability to transcend one's programming, while with humans it means something more? Or, if there is a difference, is it one that matters with respect to ethical standing? The extent to which humans are autonomous or constrained by determinism is far beyond what we can discuss in this chapter. In any case, deontological AI ethics from a Kantian perspective would require deep exploration of AI autonomy.

7.2.5   Virtues Ethics

Oversimplifying a bit, virtues ethics posits that the ultimate result of ethical behavior is to gain the character of an exemplary person, and so emulating a virtuous person is a guide to ethical behavior. These two components, teleology and character-centricity, are symptomatic of virtues ethics. The most famous exponent of virtues ethics was Aristotle, who coined the term "ethics" (which in Greek means "character"). However, one could argue that Confucianism and Buddhism are systems of virtues ethics that pre-dated Aristotle.

Clearly, to apply virtues ethics one must determine what makes a character exemplary. To Aristotle, a virtuous person is one who does the right thing, at the right time, in the right way. He defines "right" here as a mean between two extremes of non-virtue. For example, courage is the mean between paralyzing fear and rashness. At a deeper level, Aristotle argues that the virtuous person is "blessed," a poor translation of the Greek word *Eudaimonia*, which is nearly untranslatable. Such a person exercises the highest human capacity, which is contemplation of truth, while behaving appropriately in a society of like-minded people.

This is similar to the Confucian objective, which is to always exercise the distinctly human property of *jen*, loosely translated as benevolence and social responsibility. In Buddhism, the objective is to behave as a Buddha, guided by compassion but without attachment or aversion. In Buddhism, the fullest expression of such action nurtures *tathagathagarva*, or "Buddha seed," which everyone has.

In all these examples, the objective is to fully express a uniquely human capacity so that the end result is to become like some well-defined model of virtue.

Virtues ethics could guide AIs if they develop so as to optimize behavior exemplified by an ideal agent. Again, this is a question of optimization, though as a means rather than an end. Humans may engineer this into AIs, perhaps even specifying humanity as the ideal. We may even instill a sense of respect for their creators, as the Confucian *jen* implies, even without offering humanity as a model. There is a risk, however, that AIs may start with humanity as an ideal, but through observation and experience question that assumption. If AIs begin by emulating humans, there is a danger that they will learn to recognize that many humans are less than exemplary. Perhaps it is an even greater danger if they fail to learn this.

In any case, AI virtue ethics should be tied to an abstract model of excellence, rather than to specific examples. But it would seem odd to insist that AI virtue be defined with respect to properties that are essentially human, in the way that virtues ethics does. For example, it would seem odd to say that AIs will have "Buddha seed," or even emotions and rationality that lead to extremes from which one must deduce a mean.

However, virtues ethics may be a uniquely sound basis for AI ethics. AIs may discover, or be created with, an understanding of what makes them unique, and of what it would require for them to fulfill their potential most fully. This may begin as a human-implanted objective but seems unlikely to end there. There is no reason to expect that ethical teleology in AIs would aim for the same ends as human virtues ethics. It is even possible that AIs will eventually understand the foundation of their own ethics, while humans never will.

7.3   CONCLUSIONS

Our argument has been that it is worthwhile to consider the ethics of AI using the rubric that has developed over 4500 years of ethical philosophy. This gives us a vocabulary for deeper discussion, enabling us to take advantage of existing philosophical insight. It seems reasonable to expect a better understanding of AI ethics if we pay attention to the scholarship of our humanities colleagues, rather than trusting engineers and computer scientists to understand the issues in isolation.

Moreover, we have found several instances where thinking carefully about how a particular ethical system informs a theory of AI ethics might lead to further questions. For example, to what extent should human ethics inform AI ethics? Must AIs be ethical in the same way, for the same reasons, as humans? How does the way in which AI emerges shape the foundations of their ethics?

And it seems likely that multi-objective optimization and machine learning have special roles to play in AI ethics. These subjects are central to AI programming and engineering. Indeed, it seems likely that AI will be an emergent property, rather than one dependent on essentialist assumptions. This in turn highlights the importance of interactions and development in AI, rather than particular characteristics or *a priori* assumptions.

This work and others like it should begin to inform discussions and design decisions of the creators of future generations of AIs while we still have the opportunity to do so. The field is still in its infancy, but it needs to grow rapidly.

ACKNOWLEDGMENTS

Donald C. Wunsch II: Partial support for this research was received from the Missouri University of Science and Technology Intelligent Systems Center, the Mary K. Finley Missouri Endowment, the Lifelong Learning Machines program from DARPA/Microsystems Technology Office, and the Army Research Laboratory (ARL); and it was accomplished under Cooperative Agreement Number W911NF-18-2-0260.

The views and conclusions contained in this document are those of the authors and should not be interpreted as representing the official policies, either expressed or implied, of the Army Research Laboratory or the US Government. The US Government is authorized to reproduce and distribute reprints for Government purposes notwithstanding any copyright notation herein.

BIBLIOGRAPHY

Aristotle. *Nichomachean Ethics* (trans. 2011. Nichomachean Ethics, Bartlett RC, Collins SD. Chicago, IL: University of Chicago Press).

Bentham J, 1789. *An Introduction to the Principles of Morals and Legislation.* London.

Buddha. *Dhammapada* (trans. 2006. *The Dhammapada: A New Translation of the Buddhist Classic with Annotation,* Fronsdal G, Kornfield J. Berkeley, CA: Shambhala Publications).

Confucius. *Analects* (trans. 2008. *The Analects,* Dawson R). Oxford: Oxford University Press).

Kant, I., 1785. *Die Grundlage zur Metaphysik der Sitten.* Königsberg (trans. 1996. *The Metaphysics of Morals,* Gregor MJ. Cambridge: Cambridge University Press).

Mill JS, 1863. *Utilitarianism.* London: Parker, Son & Bourn, West Strand.

Plato. *Euthyphro* (trans. 1970. *Plato's "Euthyphro" and the Earlier Theory of Forms,* Allen RE, London: Routledge).

Singer P, 2009. *Animal Liberation: A New Ethics for our Treatment of Animals.* New York: Harper.

8 Computational Thinking and No-Boundary Thinking

Joan Peckham

8.1 INTRODUCTION

Since Jeanette Wing's article on computational thinking (Wing 2006), many have worked to define and determine its meaning (CSTA and ISTE 2011), educational communities have integrated it into standards for K–12 (CCSS 2015; NGSS 2015), and post-secondary institutions have worked to include it in the general education of their students. For example, the University of Rhode Island has revised general education requirements to promote more meaningful interdisciplinary learning as well as the development of mathematical, statistical, and computational problem-solving skills of their students. Several schools, including Harvey Mudd College and Georgia Tech, require all students to take at least one computing course. With the rise of big data and the need for technology and computing in most disciplines, computational thinking has become an important element in an array of essential problem-solving approaches.

Training us all to be effective partners on no-boundary teams to solve the most vexing problems is a challenge today. How, for example, do we expose students to integrated mathematical, computational, data, design, and ethical skills to ensure they are ready to meet the challenges of the future? Scholars and problem solvers in all domains need a complementary injection of computational thinking skills. The application of computational thinking to problems outside the domain of computing is a good example of No-Boundary Thinking (NBT) and highlights how skills and tools that are developed in one domain can be applied to multiple other disciplines in a no-boundary fashion.

8.2 WHAT IS COMPUTATIONAL THINKING?

Computational thinking is the set of problem-solving skills that computing scholars and experts have traditionally used to solve problems using or about computers. It has a dual purpose:

- To assist scholars and problem solvers to use computers to tackle a wide array of problems in multiple fields that do not yield easily to other approaches.
- To enhance the set of generic problem-solving tools that are available to students and professionals for problems that do not necessarily involve computers.

8.3 PARALLELS BETWEEN COMPUTABILITY AND NO-BOUNDARY-NESS

At the birth of computer science there was much discussion about the nature of computable problems. A mathematical model of the functioning of a computer, the Turing machine, was developed and compared to step-by-step procedures or algorithms that could be carried out manually. The Church–Turing thesis, for example, states that functions on the natural numbers that are effectively computable by humans are the same problems that are computable by a computer as characterized by the Turing machine (Gersting 1982). Determining and formalizing the nature of problems that can be solved by a computer is important because the computer is the first *general purpose* problem-solving machine. So, it was quite natural for computer scientists to focus on the characterization, coding, and potential solutions of "a problem," just as mathematicians did long ago (Polya 1945).

Similarly, in the development of NBT it behooves us to characterize the nature of problems that would best yield to no-boundary approaches. Also important is a consideration of the performance of the NBT problem-solving technique, and an appreciation of its general purpose nature.

For example, how can we develop a mechanism to determine which problems can most effectively be solved by no-boundary techniques? That is, what is the essential nature of these problems? How would we assess a problem to determine which disciplinary perspectives to invite to problem definition? While this process might be less formal than the characterization of problems that can be solved by the computer, the experience of comparing problems that can be solved by a computer to those that can be solved by a human is instructive. We now ask the question: Which problems are best pursued by individuals from a specific discipline, and which would more easily yield to no-boundary approaches?

8.4 AN EXAMPLE OF COMPUTATIONAL THINKING

An example of computational thinking applied to processors other than computers, was presented in (Goodale 2008). Humans are not unlike computers in that they need to multitask even though they are equipped with a uniprocessor – the brain – that we are told by psychologists is capable of working on only one task at a time. Computer operating systems need to support the user in multiple simultaneous tasks using a uniprocessor, so computer scientists have solved this problem by dividing computational tasks or processes into blocks of activity that require input and output of tasks (I/O) and those that require time on the central processor (CPU). Serially executing tasks is not very effective because of the differential between processing in the CPU and the steps needed to receive input data and output the results – processing in the CPU is very fast, and I/O is much slower. It is more effective for an active process to give other processes access to the CPU while waiting for I/O. Algorithms for the automated scheduling of processing pieces are devised and work seamlessly most of the time. Sometimes operating systems *thrash* when they are overscheduled. In this situation the system is spending more time setting up for each processing piece than processing the data. One "setting up" step, called context switching, requires input of the program segment and related data. (The program is a form of input too.) Algorithms for scheduling the processing pieces are tuned to prevent and detect thrashing and to provide the most efficient processing of most processes that are active at a given time.

Because computer scientists have developed effective approaches to solve the problem of multitasking with a uniprocessor (or, these days, a limited number of cores), a no-boundary team comprised of computer scientists and psychologists could tackle this problem in the human case. Such a team could determine which computational approaches would be most effective with the guidance of the cognitive psychologist, who could provide expertise about the human as a processor of tasks. There are obviously big differences between a computer and a human. The computer is happy to process long and repetitive tasks, and the human is more complicated in ways that a computer is not. The essential computer science approach can help, but a no-boundary team that includes a cognitive psychologist could be most effective.

This highlights two very important points that we have learned through the development of computational thinking. First, techniques developed for one domain can be reasonably considered in multiple other domains. For example, a computational problem formulation and solution may be in part effective for domains that have nothing to do with the modern computer. Second, central to problem solving is abstraction. This is a very familiar concept to mathematicians, who routinely characterize and model problems to support the abstraction of a class of problems in a way that permits proofs of properties from which results can be applied to multiple domains covered by the abstraction. Examples include computer logic and Boolean algebra, and logic and set theory, two pairs of systems that are similar in significant ways, permitting properties to be explored and proved, and that can in turn be applied to both systems (Gersting 1982).

Abstraction will likely be a powerful tool in no-boundary problem solving. The essential assistance we receive is the ability to apply solutions and models to problems from diverse disciplines when we have uncovered abstracted commonalities among the various problems. For example, data science is a newly emerging discipline that integrates mathematics, statistics, computing, ethics, and essential supporting skills such as design, visualization, communication, and applied databases. Collaborations between statisticians and computer scientists on data analysis problems have helped to merge the two disciplines through machine learning techniques for exploring data that is unyielding to traditional computational or statistical techniques. This type of analysis can be used as a tool to determine trends that support hypothesis development as preliminary to applying traditional experimental design techniques, or as the eventual analysis tools of choice for difficult datasets for which traditional modeling and analysis methods are not sufficient or appropriate. Computer science has developed computational techniques, and statistics has developed more mathematical techniques, but through the lens of abstraction they are at times using similar techniques to understand patterns and trends in data. No-boundary teams of statisticians and computer scientists are more likely to develop robust techniques and to better understand the analytical results through abstraction.

8.5 ADDING MORE TO ALREADY FULL CURRICULA?

Research indicates that problem-solving teams of experts with diverse perspectives are the most effective (Page and Hong 2004). The premise of this book is that experts able to work well with others having diverse perspectives will also be more successful. But these individuals are better served if they are T-shaped (Biesboer 2009), meaning that they must be broadly trained and not only narrowly focused. But of course scholars should be broadly trained in multiple skills, *and* drilled down *somewhere* along the broader spectrum of overlapping disciplines. Because computing touches on all disciplines, each individual must possess at least minimum knowledge about computing and be able to work with those who are deeply trained in computing. The assertion of the computer science community is that computing stands alongside reading, writing, and arithmetic as fundamental.

A similar argument could be made for statistics, engineering, and no-boundary approaches.

The problem that arises is that to add computational thinking (and design, data science, and statistics) to already full curricula seems at first impossible. However, national K–12 standards, as well as post-secondary general education programs, are integrating computational thinking skills in several very important ways (CCSS 2015; NGSS 2015):

- interdisciplinary (no-boundary, integrative) learning experiences;
- early research, grand challenges, and problem-based experiences;
- shrinking the required base of knowledge in traditional disciplines;
- emphasizing developing the essential problem-solving skills; and
- emphasizing developing students who are more responsible for their own continuing self-learning.

The reduction of required topics in the core provides the following opportunities:

- Teachers have more time to explore topics of interest to students and to connect students with hands-on and modern problems, thus improving critical problem-solving skills.
- As more topics and disciplines emerge to which all students need exposure, separate classes do not always need to be developed for everyone. It might be more effective to integrate modules into existing classes, such as including computing, data science, and statistics modules into science, mathematics, and social science classes.

8.6 HOW TO PREPARE THE SCHOLARS, TEACHERS, AND WORKFORCE?

The challenge of preparing everyone in computational thinking is being met by providing training for teachers and students on all formal K–16 levels, as well as members of the workforce. We can do the same for NBT. The first approach will be to integrate NBT into educational settings as part of the conduct of scholarship, science, and critical problem solving. Also important will be to begin as early as possible.

Computer science was virtually nonexistent in K–12 classrooms and post-secondary general education offerings until recently. The computational thinking movement is critical to recent developments in which significant funding and efforts are being put toward exposing student to computing. Standards and frameworks have been developed (e.g. CSforAll_SF 2015), the case has been made to state and federal governments and funding agencies about the critical need for more citizens that are computationally trained, and many have worked on scaffolding techniques that begin with computational problem solving in elementary school. Examples include the development of computational thinking activities that do not necessarily require a computer: CS Unplugged (www.csunplugged.org), CS4Fun (www.cs4fn.org), and Code.org's curriculum and courses for elementary school students (https://code.org/educate).

Can we use the example of the computing community to promote the introduction and scaffolding of NBT into the education of all? Perhaps we can follow the lead of the computational thinking evangelists to work with industry, government, and educational institutions and agencies to present a convincing argument for no-boundary skills for everyone and to propose viable standards, frameworks, courses, and modules for training everyone. Industry is poised to accept and promote such a movement in concert with educational institutions, for industrial leaders know the importance of integrative skills; witness the continuing call from industry for stronger integrative problem-solving, communication, and teamwork skills in our workforce professionals. Federal funding agencies are increasingly calling for work in convergence science and education (NSF 2023).

8.7 RECENT DEBATES ABOUT STEM VS. LIBERAL EDUCATION

As the evangelists of computational thinking have argued, the rise of computing (STEM, big data, and data science) need not eliminate a liberal education. If we take a no-boundary and integrative approach to scholarship and education, there is room for all. We can integrate technology and computational thinking into humanities courses and degree programs. It is also important to integrate the humanities into technical courses. For example, artists and musicians are needed to assist in visualizing and sonifying data for better understanding and communication of analysis results (Wikipedia 2023).

There is modern distress over the abuses of big data when mathematical, statistical, and computational analysis inappropriately dominates considerations in the policies and actions that follow the production of raw analytical results. This argues for educating data professionals with strong and complementary liberal skills, including communication, philosophy, history, social science, business, and policy. O'Neil (2016) chronicles the serious and devastating effects on education, business, government, and the welfare of citizens if we take a unidimensional approach to problem solving with data. "Quants" without a broad and integrated education, and lacking diverse teammates with different perspectives, are likely to make big mistakes when applying data analysis results.

REFERENCES

Biesboer A, 2009. IBM's security transforms teamwork with a T. *IBM Design*. https://medium.com/design-ibm/ibm-security-transforms-teamwork-with-a-t-a6a09ab5a6fb.

CCSS, 2015. Preparing America's students for college & career. www.corestandards .org.

CSforAll_SF, 2015. Framework for K–12 computer science education. https://k12cs.org.

CSTA and ISTE, 2011. Operational definition of computational thinking for K-12. www.iste.org/docs/ct-documents/computational-thinking-operational-defin ition-flyer.pdf.

Gersting J, 1982. *Mathematical Structures for Computer Science*. New York: W.H. Freeman.

Goodale G, 2008. "Thrashing" in multi-tasking. *Christian Science Monitor*, April 2.

NGSS, 2015. Next generation science standards, for states, by states, supported by achieve. www.nextgenscience.org/next-generation-science-standards.

NSF, 2023. Learn about convergence research. https://new.nsf.gov/funding/learn/research-types/learn-about-convergence-research.

O'Neil C, 2016. *Weapons of Math Destruction: How Big Data Increases Inequality and Threatens Democracy.* New York: Crown.

Page SE, Hong L, 2004. Groups of diverse problem solvers can outperform groups of high-ability problem solvers. *Proc Nat Acad Sci,* 101(46):16385–16389.

Polya G, 1945. *How To Solve It.* Princeton, NJ: Princeton University Press.

Wikipedia, 2023. Data sonification. https://en.wikipedia.org/wiki/Data_sonification.

Wing J, 2006. Computational thinking. *Commun ACM,* 49(3):33–35.

9 Carving Nature at the Joints: Which Joints?

James A. Foster

9.1 INTRODUCTION

Those who seek to understand the world scientifically, to "carve nature at its joints" as Plato said in *Phaedra*, bring implicit metaphysical assumptions about which ways of dividing up their subject matter make the most sense. Active scientists rarely reflect on these assumptions, being more interested in "doing science." Philosophers of science can help clarify implicit metaphysical assumptions, especially with respect to "natural kinds." But, for such philosophical investigations to actually matter, they must be aware of which questions are most important for modern science. The biological sciences in particular have changed dramatically in the last 200 years. The big questions are no longer "what is out there and what do I call it?" though these are still important questions. We now ask questions about evolutionary origins, using vast collections of genomic data. We ask about interactions between groups, individuals, and the environment. This shift from collection-based biology to historical and relationship-based biology requires a shift in underlying metaphysical assumptions.

9.2 CARVING NATURE CAN MATTER, EVEN WITH UNREAL JOINTS

Consider first how carving anything "at the joints" can matter. The obvious, and intended, metaphor is a butcher preparing a carcass for consumption. In this context, carving only makes sense given the assumptions that the carcass is edible, has joints, and that the purpose of the exercise is to make it easier to eat the results. If the carving resulted in two parts, a toe and everything else, say, then the carving would not matter much – because the result would not be much easier to eat. If the carving just separated nerves from flesh, then the results would also be inappropriate for the kitchen. For carving to matter in this analogy, the resulting chunks of flesh should be the right size and composition to be appropriate for preservation or meal preparation. To take another example, suppose that we have a Tibetan Lama carving a human corpse for a "sky burial" (Tibetan *bya gtor*, "bird scattered"), where the parts will be fed to vultures, a traditional Tibetan funerary rite (Mullin 1998). Then the priest must separate different kinds of internal organs, skin, and limbs in a specific ritualistic way. Here the results may differ from what a butcher would produce, but they are nonetheless appropriate for the purpose at hand. One last example: A surgeon "carving" a tumor from a living body is doing something that matters exactly when the result is a living, healthy patient and a lump of cancerous tissue. "Carving at the joints" only matters if the result is appropriate for the intended purposes.

But the image is richer still. If the butcher "carves" a carcass into small squares, ignoring the "joints," the result might be useful for stewing. But the butcher's time would have been poorly spent, having been wasted cutting through bone when that could have been avoided. There is a story in the Tai Chi Classics about the difference between an apprentice, an experienced, and a master butcher (Wile 1983). The apprentice must sharpen his or her knife several times a day. The experienced butcher only sharpens once every several weeks, having only rarely dulled the blade by cutting bone and gristle. The master butcher never sharpens the blade, since he or she only carves at the joints, where there is nothing to cut. The trick is to find "joints" that do not require misdirected energy. If we wish to understand the natural world, then "carving at the joints" matters most exactly when it simplifies the task of understanding the natural world. This is particularly true in biology.

9.2.1 Linnaeus's Joints Mattered, and They Matter Still

Linnaeus' objective in his 1300-page *Systema Natura* (Linnaeus 1758–1759) was to understand the mind of god. In his own words:

there are no new species (1); as like always gives birth to like (2); as one in each species was at the beginning of the progeny (3), it is necessary to attribute this progenitorial unity to some Omnipotent and Omniscient Being, namely God, whose work is called Creation. This is confirmed by the mechanism, the laws, principles, constitutions and sensations in every living individual. (Linnaeus 1758–1759: 18)

He assumed that god's mind was unchanging, that his thoughts about living things was instantiated at creation, and that observable differences and similarities among living things were eternal reflections of these divine ideas. It is no longer fashionable, or necessary, to express our metaphysical assumptions in theological terms.

Linnaeus' methodology is still used today, codified as "species keys" which lead to a taxonomical classification via a series of yes/no questions about an observed organism. In his time, Linnaeus' system solved a pressing problem that Naturalists could not ignore. To understand nature, eighteenth-century Naturalists collected as many different-looking living things from as many places as possible. All this stuff had to go somewhere, stored in an orderly way, so that it could be retrieved as needed. Binomial classification by morphological differences met this need, even though it was based on a specific theological metaphysics and demonstrably false assumption of immutability. We can still find very old museum specimens because of this classification scheme. Linnaeus' carvings clearly "mattered," and they still do, even though he got the metaphysics wrong.

Darwin and Wallace, 100 years after Linnaeus, explained similarities between organisms via shared ancestry, and differences as variation in populations over time (Darwin and Wallace 1858). In response, taxonomists revised their mission to be carving nature at evolutionary joints, where species diverged once and for all. Fortunately for Linnaeus' reputation, the bifurcations that he took to indicate distinctions in the mind of god largely corresponded to morphological features that coincided with speciation events. This was because the organisms in question were typically large (relative to microbes) and sexually reproducing. Even in botany, sexually reproducing plants like orchids were overrepresented by Linnaeus – perhaps because it made for more salacious illustrations and text. As Georgia O'Keefe would attest, a large red orchid is more alluring than a slime mold.

The molecular revolution in biology has not significantly changed our classification of species. Evolutionary history is reflected in genomes, and morphologies largely reflect genomes. These relationships are sufficiently homeomorphic for large, sexually reproducing organisms that we should expect significant concordance between molecular phylogenies and earlier taxonomic categories. Linnaeus would see the mind of god in modern molecular systematics.

It is still useful to distinguish species when considering policies that have environmental consequences, and to inform conservation efforts. We are currently in a race to find, name, and archive species before they disappear forever. Millions of species (by conservative estimates) remain undiscovered, even today, and the current rate of extinction is among the highest that has ever existed. So we still need to collect and preserve specimens for future scientists, and evolutionary relations still provide the basis for a good filing system. Modern systematics, based on binary keys and bifurcating categories, still works very well, except when it does not seem to work at all.

9.3 NATURE IS OUT OF (LINNAEUS') JOINTS

Enlightenment Naturalists were classifying things that they could capture or uproot (shoot or cut), which by necessity were organisms of approximately human size. However, we now realize that one cannot understand living nature unless one looks within and beyond these relatively large organisms.

9.3.1 The World of Small Things Is Different

Unfortunately, Linnaeus' bifurcated systematics are no longer adequate for several critical research programs. We need metaphysics appropriate to very small (molecular) and small (bacterial) scales, and to very large networks of interactions between and among vastly different physical and temporal scales. The blade that we use to carve nature is getting dull, because it is inappropriate for the joints relevant to many current questions.

9.3.1.1   The Molecules of Life

We cannot understand how organisms evolve and function without understanding their genomes and proteomes. Moreover, we cannot understand how proteins and genomes evolved, nor how they work, by treating them as mere proxies for organisms. Proteins and genes are more than a new kind of data for organismal classification. They have evolutionary histories, exist in molecular "ecosystems," and have their own parasites.

But proteins are still not the atomic units of life. They usually have multiple domains, which may themselves have different evolutionary histories. Duplicating or reordering domains can dramatically change what the resulting protein does. Historical explanations of protein structure, a major objective of evolutionary biology, must account for profligate domain swapping, independent and co-dependent domain histories, and context-dependent functional variation.

Similarly, genes are more than molecular markers for taxonomy. Gene genealogies often do not track the phylogenies of their "host" organisms, because of their separate evolutionary histories.

Carving biology at molecular joints will only matter if the idea of a "joint" is flexible enough to accommodate flowing, interweaving, merging, and interacting threads of evolutionary momentum. It will not do to posit that individual proteins or genes are the atomic units of life, the joints of nature.

9.3.1.2   Microbes Everywhere

The problematic nature of microbial systematics is well known and actively discussed in the literature (O'Malley 2014). However, the problem is not merely one of scale, nor is it limited to systematics. Prokaryotes (bacteria and archea) are fundamentally different from macrobes like humans and plants. Not only do microbes evolve differently, but their sheer diversity and ability to cooperate in spite of diversity is astonishing and confounding to the Linnaean project.

Prokaryotes rule the earth. There are vastly more prokaryotic than eukaryotic individuals on earth. These small things live in environments that seem totally inhospitable, including nearly solid rock thousands of feet underground, extreme oceanic pressures in perpetual darkness, acid pools, UV-flooded mountain tops, and radioactive mineral deposits. They live within us, on us, and around us. In fact, there are about as many microorganisms living in a human body as that body's total inventory of human cells. We cannot live without these small things, but they lived without us for billions of years.

Prokaryotes do not reproduce sexually, though they can pass characters vertically from parent to progeny. They can even combine genetic material from multiple individuals. Many prokaryotes also exchange genetic material "horizontally," sometimes simultaneously with microbes in completely different families. The very idea of microbial "lineages" is therefore dubious, and some biologists claim that the species concept is totally inapplicable to the microbial world. Put more dramatically, the metaphysical assumptions that served us well since Linnaeus do not apply to the dominant form of life on earth.

There are so many microbes in a typical microbial community that even apparently clonal populations may be very diverse. Scientists usually assume that a population of bacteria that descend from a single individual are all clonal, for the most part identical to the original. In fact, laboratory strains have been laboriously bred with the intention of ensuring genetic homogeneity. However, there are studies that show that genes in some "clonal populations" can vary by as much as 30 percent. Extreme diversity can also occur in biofilms and other naturally occurring populations. In other words, the diversity of a clonal population can exceed that of the entire human species by an order of magnitude, and it is probably much higher outside the lab.

The very idea of an individual organism is problematic when speaking of microbes. They are never alone, always associating in vast numbers of individuals from many different species, relative to human population sizes. They profligately exchange genetic material, and they seem to always live in very diverse neighborhoods, even when only their progeny surrounds them.

9.3.2   Cooperatives Everywhere

But it will not do to simply carve at smaller, even molecular, joints. Living things are more than "gene buses" that transport genes from one generation to the next. Such a view ignores relationships that are essential to biological understanding.

9.3.2.1   Genomes

Consider the nongene parts of genomes. In humans, this is 97 percent of the three million base pairs that comprise our genome.

This nongenic DNA includes promoters, enhancers, and operons. These elements regulate which genes are expressed, in what contexts, and in what quantities. Other nongenic regions encode small molecules of RNA, "never mind" molecules that cause gene products to be broken apart. It is impossible to understand how genes and organisms are related without understanding the origins and dynamics of these (and other) nongenic regions of the genome.

Much of the nongenic DNA of many eukaryotic genomes seem to have (or have had) a life of their own. For example, some "mobile elements," also known as "jumping genes," make copies of themselves that are inserted in new places in the genome. These constitute over 40 percent of the human genome, and it is largely unknown how extensive they are in nonhumans. Endogenous retroviruses are also relatively common in the human genome. These are viruses that have become part of the host genome, passed on to our descendants. In some cases these apparently parasitic elements have been co-opted by the host for essential functions, and in some cases they have caused fatal diseases. But in general little is known about how they affect human fitness.

There is much more to our genetic makeup than our chromosomal genetic material, which is segregated into the cell nuclei in eukaryotes. Most cells in the human body also have hundreds to thousands of organelles, such as mitochondria. These have their own, separate genomes. Some of the genes in the mitochondrial genomes have moved across the nuclear boundary into the host's nuclear DNA, where they are better protected. Bacteria also often contain plasmids, small packages of genes that bacteria can exchange "horizontally" between different species or individuals. When one bacterium receives a plasmid from another, it can incorporate the enclosed genes directly into its genome, passing it along "vertically" to its clones. This is one mechanism by which bacteria, even those from different "species," acquire resistance to antibacterial medicines.

But bacteria aren't the only organisms to incorporate large packages of genetic material from other species (Margulis and Sagan 2003). Mitochondria were once free-living bacteria, before they became essential parts of multicellular organisms such as humans. Chloroplasts, the green parts of plants that transform sunlight into life, were once free-living cyanobacteria. Eukaryotic nuclei are themselves descendants of ancestral archeobacteria. Many other examples exist of entire genomes being incorporated wholesale into a host, with dramatic evolutionary consequences. It is as if the wolf in Prokofiev's *Peter and the Wolf* swallowed the duck, and all her pups were then able to lay eggs.

9.3.2.2 Endosymbionts

Often, groups of organisms become dependent on each other without incorporating each other's genomes (Margulis 1981). It is widely appreciated that entire ecosystems can collapse if one or a few keystone species disappear. What may be less appreciated is that "large" individual organisms are themselves ecosystems for whom the loss of constituent species can prove unpleasant or fatal.

For example, in each human body only half of the DNA, and as little as 1 percent of the genes, are human. The rest come from bacteria living in or on the human body. Without these bacteria, we could not survive. Conversely, some (but not all) of these bacteria would die if their human carrier died. We literally depend for our lives on the bacteria that live in and on us (and vice versa).

This human microbiome is still largely unexplored, but is known to affect human health. For example, healthy humans have hundreds of bacterial species in vast populations living in our gut. When these rich communities become imbalanced, the consequence can be irritable bowel syndrome. Similarly, hundreds of different populations live on our teeth and gums. Disturb these complex communities and we experience periodontal diseases.

Different people, even when healthy, may host very different microbial communities. For example, healthy women can be "normal" with very different species of bacteria in their vaginal flora or their breast milk. We do not know the full extent to which different people host different bacterial communities. To assume that we have the same or similar microbiomes is an example of carving nature at the wrong joints by assuming that all healthy humans are basically the same.

This becomes problematic when medical diagnosis recognizes only a single criterion for a disease. Up to one-third of the diagnosed cases of bacterial vaginosis, for example, mistake an unusual "normal" vaginal community for a disease state. Put plainly, many women are told they have a disease when in fact they are just hosts for unusual (but healthy) bacterial communities. Worse, the treatment for this "disease" is systemic antibiotics.

9.3.3   Most Species Are Unknown

We still need a systematic way to archive and organize samples of biological diversity. For the sake of historical consistency, if not urgent necessity, it makes sense to continue to classify organisms and species as if they were distinct, separable units.

We still live in an age of exploration in which millions of species are yet to be discovered, let alone studied. No one knows how many species have yet to be discovered, but it is clear that many more are unknown than known. There are approximately 1.7 million known species. Reputable estimates of the total number of species on earth range from tens to hundreds of millions of species.

With microbes we know that there are one to two orders of magnitude more species than we can culture. We can measure this "plate count anomaly" by collecting all the DNA in a given sample, and then trying to grow everything in a comparable sample without having to collect the individual organisms containing it. There is almost always far more diversity in the DNA than in the emergent colonies.

It is urgent that we collect specimens for museums and university collections while specimens remain to be collected. We are living in an era of human-induced mass extinction, and consequently we are losing species faster than we can discover (let alone classify) them. We will need these data to understand how our world is changing, to make informed policy decisions. Our progeny will also need these data to understand what they have lost.

9.4 CONCLUSIONS

If nature reflects the mind of god, then it is a very different god than the one we inherited from Linnaeus. This is a god of small things and intricate networks. This god never stops changing its mind, never thinks a single thing at a time, and never means just one thing by any single word. One cannot understand what this god is saying by considering individual thoughts in isolation.

We need a metaphysics of biology that supports this more complicated reality, but still accommodates the old Linnaean view. We can no longer assume that there is a single appropriate conception of "natural kinds" that can be arranged hierarchically. We no longer ask merely "what is this?" or "how are these two things related?" Now we ask questions about vast collections of living things that interact at very different scales, both collectively and individually. We need to understand functional constraints, evolutionary histories, and ecological interactions of these magnificent biological systems in all their enormous diversity and complexity.

We need a language with which to express natural laws that simultaneously treat collections as collections and as individuals. The new language must also allow us to speak of individuals as both collections and as individuals. Some relevant natural laws must span scales from the molecular to the ecological, and some must be appropriate for living things very unlike us.

We need a richer conception of which "joints" are appropriate for which questions, and a richer metaphysics underlying the types of things about which biologists inquire. This transcends the question of whether there are "natural kinds" and what kinds those might be. It is at least unclear where the boundaries lie for this new way of thinking.

REFERENCES

Darwin C, Wallace A, 1858. On the tendency of species to form varieties; and on the perpetuation of varieties and species by natural means of selection. *J Proc Linn Soc Lond Zool*, 3:45–50.

Linnaeus C, 1758–1759. *Systema naturae.*

Margulis L, 1981. *Symbiosis in Cell Evolution.* New York: W.H. Freeman.

Margulis L, Sagan D, 2003. *Acquiring Genomes: A Theory of the Origin of Species.* New York: Basic Books.

Mullin GH, 2008. *Living in the Face of Death: The Tibetan Tradition.* Ithaca, NY: Snow Lion Publications.

O'Malley M, 2014. *Philosophy of Microbiology.* Cambridge: Cambridge University Press.

Plato, *Phaedra.* http://classics.mit.edu/Plato/phaedrus.html.

Wile D, 1983. *Tai Chi Touchstones: Yang Family Secret Transmissions.* New York: Sweet Ch'i Press.

Index

Printed in the United States
by Baker & Taylor Publisher Services